Automatic Parallelization

An Overview of Fundamental Compiler Techniques

Synthesis Lectures on Computer Architecture

Editor
Mark D. Hill, *University of Wisconsin*

Synthesis Lectures on Computer Architecture publishes 50- to 100-page publications on topics pertaining to the science and art of designing, analyzing, selecting and interconnecting hardware components to create computers that meet functional, performance and cost goals. The scope will largely follow the purview of premier computer architecture conferences, such as ISCA, HPCA, MICRO, and ASPLOS.

Processor Microarchitecture: An Implementation Perspective
Antonio González, Fernando Latorre, and Grigorios Magklis
2010

Transactional Memory, 2nd edition
Tim Harris, James Larus, and Ravi Rajwar
2010

Computer Architecture Performance Evaluation Methods
Lieven Eeckhout
2010

Introduction to Reconfigurable Supercomputing
Marco Lanzagorta, Stephen Bique, and Robert Rosenberg
2009

On-Chip Networks
Natalie Enright Jerger and Li-Shiuan Peh
2009

The Memory System: You Can't Avoid It, You Can't Ignore It, You Can't Fake It
Bruce Jacob
2009

Fault Tolerant Computer Architecture
Daniel J. Sorin
2009

The Datacenter as a Computer: An Introduction to the Design of Warehouse-Scale
Machinesfree access
Luiz André Barroso and Urs Hölzle
2009

Computer Architecture Techniques for Power-Efficiency
Stefanos Kaxiras and Margaret Martonosi
2008

Chip Multiprocessor Architecture: Techniques to Improve Throughput and Latency
Kunle Olukotun, Lance Hammond, and James Laudon
2007

Transactional Memory
James R. Larus and Ravi Rajwar
2006

Quantum Computing for Computer Architects
Tzvetan S. Metodi and Frederic T. Chong
2006

Automatic Parallelization: An Overview of Fundamental Compiler Techniques

Samuel P. Midkiff

ISBN: 978-3-031-00608-1   paperback
ISBN: 978-3-031-01736-0   ebook

DOI 10.1007/978-3-031-01736-0

A Publication in the Springer series
SYNTHESIS LECTURES ON ADVANCES IN AUTOMOTIVE TECHNOLOGY

Lecture #19
Series Editor: Mark D. Hill, *University of Wisconsin*
Series ISSN
Synthesis Lectures on Computer Architecture
Print 1935-3235   Electronic 1935-3243

Automatic Parallelization

An Overview of Fundamental Compiler Techniques

Samuel P. Midkiff
Purdue University

SYNTHESIS LECTURES ON COMPUTER ARCHITECTURE #19

ABSTRACT

Compiling for parallelism is a longstanding topic of compiler research. This book describes the fundamental principles of compiling "regular" numerical programs for parallelism. We begin with an explanation of analyses that allow a compiler to understand the interaction of data reads and writes in different statements and loop iterations during program execution. These analyses include dependence analysis, use-def analysis and pointer analysis. Next, we describe how the results of these analyses are used to enable transformations that make loops more amenable to parallelization, and discuss transformations that expose parallelism to target shared memory multicore and vector processors. We then discuss some problems that arise when parallelizing programs for execution on distributed memory machines. Finally, we conclude with an overview of solving Diophantine equations and suggestions for further readings in the topics of this book to enable the interested reader to delve deeper into the field.

KEYWORDS

compilers, automatic parallelization, data dependence analysis, data flow analysis, intermediate representations, transformations, optimization, shared memory, distributed memory

Contents

Preface

This lecture describes the fundamental principles of compiling "regular" (typically numerical) programs with the goal of extracting and exploiting parallelism. This class of programs covers much of the important domain of dense linear algebra, stencil computations, and so forth. The topics covered include a high-level overview of compilers, analyses performed by the compiler to understand the flow of data through a program and transformations to extract parallelism and otherwise increase the performance of the program. Although the main focus of this lecture is on compiling for shared memory machines, we discuss in detail the issues raised when targeting distributed memory machines. An understanding of all of these techniques will allow computer architects to better leverage the capabilities of compilers and to understand their limitations. The material in this lecture should be accessible to an upper-level undergraduate in Computer Science or Engineering with experience in programming and a basic understanding of computer systems. We have tried to make this lecture self-contained and do not assume any prior knowledge of compiler internals on the part of the reader.

We begin our lecture in Chapter 1 with an overview of what allows a program to execute in parallel, parallel execution on shared and distributed memory architectures and compiler support for parallel machines.

In Chapter 2, we describe analyses that allow the compiler to understand the flow of data through a program: data flow analysis, abstract interpretation and dependence analysis, which uses array subscript information to develop more accurate information. One important result of these analyses is the ability to determine, at compile time, when the equivalent of read-after-write, write-after-read and write-after-write hazards occur in programs. The increasingly important topic of how to analyze explicitly parallel programs is also covered. In Chapter 3, we show how the results of these analyses can be used to detect parallelism in programs, including general parallelism, vector parallelism and instruction level parallelism. The issues raised by *while* loops and recursion are also discussed. Because it is often the case that programs must be transformed before parallelism can be uncovered and exploited, Chapter 4 covers transformations that eliminate and enforce dependences (i.e., hazards), allowing parallelism to be realized.

The analyses that allow parallelism to be detected and exploited also allow the compiler to determine when other transformations that benefit performance can be used. These are covered in Chapter 5. In addition to loop interchange and tiling, which are important transformations that improve locality, we discuss transformations to block loops (useful in both tiling and targeting vector hardware), loop unrolling (which exposes instruction level parallelism and larger branch-free regions of code to target with other transformations), fusion and fission, (which allow loops to be merged and can benefit locality), and other transformations.

Chapter 6 looks at problems presented by distributed memory machines, in particular the problems of how to represent the distribution of data across the different processes executing a program and communication selection.

In Chapter 7, we go into more detail on how to solve Diophantine equations, i.e., equations whose domain, range and coefficients are integers. Solving these equations is an important technique for dependence analysis and performing unimodular transformations. Finally, in Chapter 8 we present a guide to further readings for the reader interested in exploring in depth topics raised earlier in the book.

This lecture would not have been finished without the support of numerous people. I would like to thank my wife, Laura, and my three sons, Danny, Jack and Nathan, for their love and support during this project. Projects often take longer than we think they will when we first embark upon them, and every hour spent on the book was one less hour to spend with them. I would also like to thank my mother and late father whose patience and support got me to a point in life where writing a book like this is possible.

Several people have offered invaluable feedback that has made this lecture much better than it would have been. In particular, Gagan Gupta, a PhD student of Gurindar Sohi at Wisconsin, did an extremely careful reading of an early draft and made excellent suggestions. Keshav Pingali, an anonymous reviewer, and Mark Hill did a reading of the "final" draft and their suggestions led to the inclusion of new material and a much improved and more complete book.

Samuel P. Midkiff
January 2012

CHAPTER 1

Introduction and overview

Although interest in compiling for parallel architectures has exploded in the last decade with the widespread commercial availability of multicore processors [106, 110], the field has a long history, and predates the advent of the multicore processor by decades. Research has focused on several goals, the most ambitious being support for automatic parallelization. The goal of automatic parallelization is for a compiler to take a "dusty deck" (an unaltered and unannotated sequential program) and to compile it into a parallel program. Although successful automatic parallelization has been achieved for some programs, most attention in recent years has been given to compiler support for languages like OpenMP. OpenMP allows the programmer to tell the compiler about parallel regions of a program and to provide other hints, and then have the compiler generate efficient code for the program. In both the parallelization of dusty-deck programs and support for languages with parallel annotations, the focus has been on regular and dense array-based numerical programs. In this monograph we focus largely on the first goal of dust-deck parallelization as it allows the development and explanation of much of the knowledge, and many of the fundamental techniques, required to achieve the second goal. Techniques that target non-numerical and irregular numerical programs will be discussed briefly in Section 8.9.

The first real *dependence*-based auto-parallelizing compiler (hereafter simply referred to as a *parallelizing compiler* or simply a compiler) was the Parafrase [132] system developed at the University of Illinois. The code and techniques in this compiler led, in turn, to the Ptool compiler [8] at Rice University and the PTRAN project [6] at the IBM T.J. Watson Research Center. Many of the techniques described in this work were present, either full-fledged or in a more primitive form, in these early compilers. It is interesting that a large part of the early motivation for Parafrase was to allow the study of how to develop architectures to exploit the latent parallelism in off-the-shelf "dusty deck" programs [129], as well as how to automatically compile for the Cray-1 [204] machines that dominated supercomputing at the time.

Motivated in part by the early work in parallelizing compilers, a variety of early shared memory mini-supercomputers were developed, chief among them the Alliant FX series [228], Denelcor HEP [211], and Convex [47] machines. These machines were characterized by shared memory processors, interesting architectural innovations, and overall good performance on well-tuned benchmarks. With the rapid progress in microprocessor capabilities, whose performance followed Moore's law, it became increasingly difficult for the custom processors in the mini-supercomputers to offer a significant performance advantage over cheaper microprocessor based machines. The triumph of the "killer micro" [107] was at hand, and with it the decline of both shared memory machines designed for high performance numerical computing and large-scale parallelization efforts targeting shared memory machines. Thus, a research area begun to understand the architectural implications of latent parallelism in programs was significantly, and negatively, impacted by advances in computer architecture, at least until those advances led to shared memory multicore processors.

Most of the remainder of this lecture focuses on regular numerical programs targeting a shared memory programming model, where the data of each thread are accessible to other threads, and communication is via loads and stores to memory. This is in contrast to a distributed address space programming model, where different processes communicate via inter-process messages, which is discussed in Chapter 6. Although the question of which of the models is "best" is open, the wide availability of shared memory multicore processors makes programming models that target them a research area with both intellectual depth and great practical importance.

1.1   PARALLELISM AND INDEPENDENCE

When a program executes in parallel, different parts of the program are spread across multiple processors or cores and allowed to execute simultaneously. Note that in a modern machine, different processors, and cores on a processor, may execute at different rates and with different capabilities. While this can be because of the result of differences in the hardware executing the program, i.e., different processors on a parallel machine may have different clock rates or functional units, even on identical hardware the speed at which a program region executes may vary widely on different processors or cores. Data and code cache misses, memory bus contention, O/S processes being invoked, and different branches being taken in the program, among other reasons, can cause program performance to vary.

Because of this, if the part of a program executing on one thread produces (consumes) data read by (produced by) another thread, no guarantees can be made about the rate of execution the data producer relative to the data consumer. This is shown graphically in Figure 1.1(b) and (c), which show the parallel loop of Figure 1.1(a) executing in parallel. If some processors are delayed (e.g., the one executing iteration i = 0) then read a[1] in iteration i = 2 may not get the correct values.

The task of a parallelizing compiler is four-fold. First, it must recognize when different regions of a program *may* produce, or consume, data produced by other regions. Second, it should try to transform the program to reduce the amount of interaction among different parts of the program that may prevent parallelization. Third, it should perform additional optimizations on the programs so that the sequential threads of execution in the parallelized program execute efficiently. Finally, it should produce a parallel program. We now describe with a concrete example these different tasks.

1.2   PARALLEL EXECUTION

The parallel execution of a program can be viewed in at least two dimensions. The first dimension differentiates between parallelism that targets shared and distributed memory systems, and the second dimension differentiates between the structure of the parallelism that is being exploited, independent of the target machine. We will first discuss the differences between targeting shared

```
for (i = 0; i < n; i ++) {
    a[i] = ...
    ...a[i − 1]
}
```

(a) A loop with data produced and consumed in different iterations.

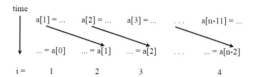

(b) The iterations execute such that reads of a consume the correct value.

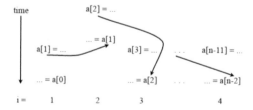

(b) The iterations execute such that reads of a consume incorrect values.

Figure 1.1: An example of how independence of references is necessary for parallelism.

and distributed memory computers, and then will discuss the three main structures or forms of parallelism.

1.2.1   SHARED AND DISTRIBUTED MEMORY PARALLELISM

The grossest difference between programs executing on shared and distributed memory is how communicating threads (or processes) communicate. In a shared memory environment, threads communicate with one another via reads and writes to shared memory. This is shown in Figure 1.2(a) and (b). In a distributed memory environment, processes communicate with one another via messages from one processor to another. This is shown abstractly in Figure 1.3(a) and (b). In both cases the simple loop of Figure 1.1(a) has been split into two loops, with each loop parallelized.

Shared memory execution

In the shared memory version of the program, each thread executes a subset of the iteration space of a parallel loop. A loop's iteration space is simply the Cartesian space defined by the bounds of the loop, and instantiated at runtime as the loop executes. In Figure 1.4(a) and (b) a singly nested loop and its iteration space are shown, and parts (c) and (d) show the same for a doubly nested loop. Each point, or node, in the iteration space represents all of the statements occurring in corresponding iteration of the loop. Graphical representations of the iteration spaces (such as what is shown in Figure 1.5) are called *Iteration Space Diagrams* and are often annotated with additional information, including constraints on how the iterations can execute (see Section 2.3).

Parallel parts of the program are placed into a worker function that has, as parameters, all of the variables accessed in the loop as well as the bounds of the loop executed within a thread. This transformation is sometimes called *outlining* in the literature [50, 230]. The pseudo-code for this is shown in Figure 1.5(a). Each parallel construct (in this case a loop) is terminated, by default, with a *barrier*. The barrier ensures that all threads executing the parallel construct finish before any part of the program after the parallel construct is executed. This barrier ensures that the results of memory operations in the parallel construct are available for all threads that might execute code after the parallel construct.

As shown in Figure 1.5(b), for execution on a shared memory machine the sequential code preceding the parallel loop is executed by a single thread. When the parallel loop is encountered, a call is made to the function `enqueueWork` that (i) places an entry in a work queue that contains a pointer to the outlined function and the argument list for the outlined function, and (ii) wakes up worker threads from a pool to execute the differences instances of the outlined function. Each thread grabs work from the work queue until it is empty, executes the barrier, and then puts itself to sleep. The initial, or master thread, then continues executes. The process continues with the call to `enqueueWork( )` to execute the outlined function for the loop containing $S2$.

Values used by $S2$ are made available to $S2$ by the stores executing in $S1$. There is generally no guarantee that the thread executing iteration $i = v$ of some loop is the same as the thread executing some iteration $j = v$ of another loop. However, values written and read are available to all threads through the shared memory.

Distributed memory

In the distributed memory version of the program, the data being operated on is distributed across the processes executing the parallel region, as shown in Figure 1.2(a). Each process also contains a copy of the entire program, but the bounds of parallel loops have been shrunk to correspond to the

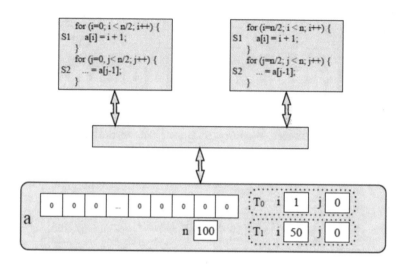

(a) A shared memory program just as the first loop begins executing.

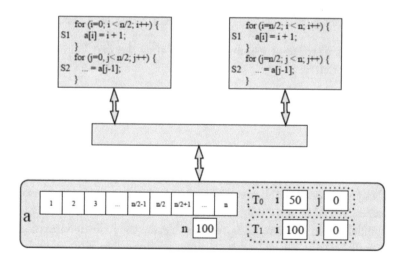

(b) A shared memory program after the first loop has executed.

Figure 1.2: An example of data sharing in a shared memory program.

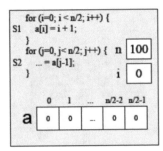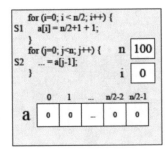

(a) A distributed memory program just as the first loop begins executing.

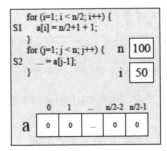

(b) A distributed memory program after the first loop has executed.

Figure 1.3: An example of data and work distribution and in a distributed memory program.

iterations of the parallel loop executed by this process. Sequential parts of the program are executed, by default, on a single processor. Each process also has access to a logical processor id (i.e., pid in the figure) that can be tested to allow only certain processors to execute selected code.

When a parallel region is entered, all processes execute the indicated iterations of the parallel loop. When the parallel loop is finished, each processor is responsible for communicating (via a *send*, in this case) the data it has produced that is needed by other processes. Each process, in turn, is required to accept (via a *receive*, in this case) the data that it needs and that is sent by other processes. The send and receive operations not only maintain a consistent view of memory for the different processes, but it also force the use of the sent data in *S2* to wait until the data is available. Thus, the communication also serves the purpose of the barrier operation in the shared memory program by forcing an ordering between accesses. The state of memory after the sends and receives have executed is shown in Figure 1.3(b).

```
for (int i = 0; i < n; i++) {
    p = p + a[i] * b[i];
    s = s + a[i] + b[i];
}
```

(a) A singly nested loop nest.

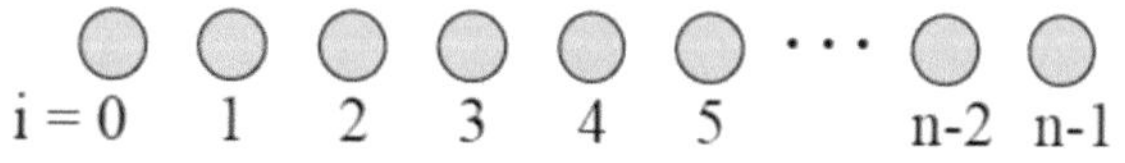

(b) The one-dimensional iteration space of the loop of (a).

```
for (int i = 0; i < n; i++) {
    for (int j = 0; j < n; i++) {
        p[i] = p + a[i] * b[i];
        s[i] = s + a[i] + b[i];
    }
}
```

(c) A doubly nested loop nest.

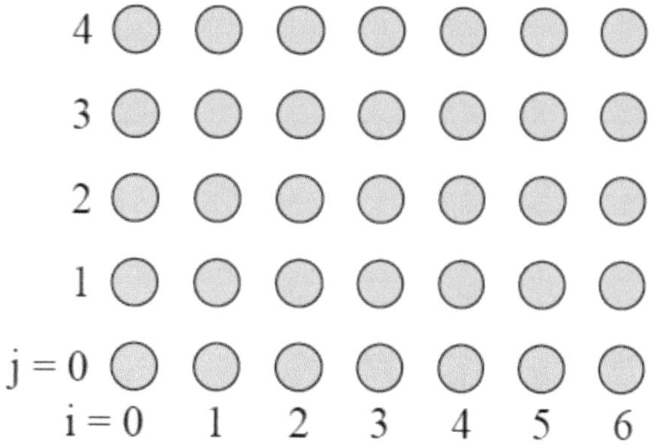

(d) The two-dimensional iteration space of the loop of (c).

Figure 1.4: An example of one and two-dimensional iteration spaces.

1.2.2 STRUCTURES OF PARALLEL COMPUTATION

Three basic structures, or forms, of parallel computation encompass the parallelism needed by most, if not all, applications. These three forms are *Single Instruction, Multiple Data* (or SIMD), *Single Program Multiple Data* (or SPMD), and *Multiple Instruction Multiple Data* (or MIMD).

SIMD execution is characterized by multiple processors or functional units executing the same instruction simultaneously on different data elements. This led to simpler hardware by allowing a single control unit to control many functional units. This is the form of computation used, for example, on the early Illiac machines [34], the Goodyear MPP [25], and the first Thinking Machines computer, the TM-1 [102]. In these machines, a bit for each processing element could be set to signal the processor to act on its data, or do nothing for this instruction. This allowed a limited form of customization of execution on each processor. Vector processing, using instruction sets such as those supported by Intel's Streaming SIMD Extensions (SSE) [215, 216] or the Power Altivec [11, 12] instructions, provide a form of SIMD execution, since they apply a single instruction or operation to the elements of a vector. Graphics Processing Units, or GPUs [13, 111, 164], which have become popular as accelerators for numerical applications [38, 60, 142, 149, 205, 227], are another example of n = 100;

...

```
for (i = 0; i < n; i ++) {
S1      a[i] = i + 1;
}
      for (j = 0, j < n; j ++) {
S2      ... = a[j − 1];
}
```

(a) A program with two parallelizable loops.

n = 100;

...

enqueueWork(*loop1, a, &n, 1, n, numThreads);
enqueueWork(*loop2, a, &n, 1, n, numThreads);

...

```
void loop1(double a[ ], int * n, int lb, int ub,
    int numThreads) {
    for (i = 0; i < n/2; i ++){
S1      a[i] = i + 1;
    }
}
void loop2(double a[ ], int * n, int lb, int ub) {
    for (j = 0, j < n/2; j ++ ){
S2      ... = a[j − 1];
    }
}
```

(b) The program after outlining.

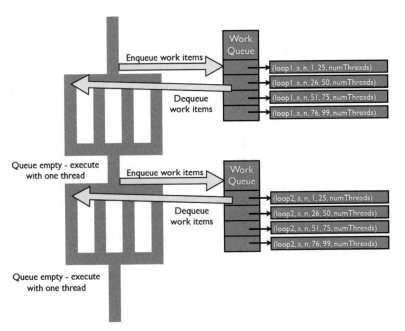

(a) Scheduling and execution of a shared memory program.

Figure 1.5: Execution of a shared memory program.

a modern SIMD machines. The deficiency of this form of parallel execution is that all computation must proceed in lock step, and processing elements not currently executing a particular operation are *not* free to perform other operations. Thus, at an `if`, the processing elements that need to execute the true branch might execute first while the processing elements needing to execute the false branch would be idle. Next, the processing elements that executed the part of the program in the true branch would be idle while the false branch processing elements execute.

The synchronous execution of instructions on these machines has implications for parallelization. Thus, in the example of Figure 1.1, the dependences across iterations would be enforced by the synchronous execution on the machine. Properly designed, these machines exhibit simpler control structures than equivalently powerful machines that execute the next two forms of parallelism.

SPMD execution consists of multiple copies of the same program each independently (i.e., not in lock step) executing against its own data set. The distributed memory execution shown in Figure 1.3(a) and (b) is an example of an SPMD execution—each process contains its own version of the program executing on its data. Shared memory programs executing many identical threads can also be thought of as SPMD programs, although as commonly used the term often refers (perhaps too narrowly) to distributed memory programs. For some analyses, the SPMD structure of the program simplifies analysis of the program. The IBM Blue Gene [29, 162] machines, clusters of workstations [173] and Beowulf clusters [27] are examples of systems that execute SPMD programs.

MIMD programs have different code executing on different processors, each operating on their own data. A multithreaded program where not all threads are identical are an example of this. The importance of MIMD programming has increased dramatically in recent years with the dominance of multicore processors, such as the Intel Sandy Bridge (see, for example, [207], IBM Power 7 [179, 180], and AMD [14] processors, as well as numerous other processors in the general purpose, embedded and high-performance computing markets.

Finally, the different forms of computation can be mapped onto one another. Thus, by merging all of the different programs in an MIMD program into one program, with control flow picking which processor executes which program, an MIMD program can be executed as an SPMD program. An SPMD program can be executed on an SIMD machine at the expense of many processing elements being idle, such as when one processing element's program performs extra iterations, and so forth.

1.3   COMPILER FUNDAMENTALS

This section describes the fundamental compiler data structures that are used in many, if not most, compilers. The analyses and transformations whose focus is program parallelization often utilize these structures in their work. As well, many papers that treat the topics of this synthesis lecture in depth (see Chapter 8) often assume a knowledge of these data structures and related compiler basics.

1.3.1   COMPILER PHASES

The high level flow of a compiler is shown in Figure 1.6. Inputs and intermediate files are shown as rounded boxes, and actual phases of the compiler are shown as boxes.

```
void foo(intn; int * p) {
    int i;
    for (i = 0; i < n; i++){
        int a = 0;
        a[i] = *p + a ++;
    }
}
```

(a) A program unit to be compiled.

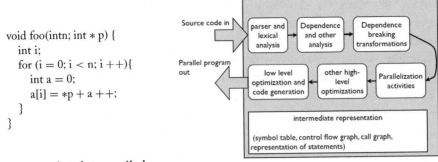

(b) A high-level view of a compiler.

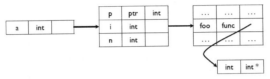

(c) The symbol table for the program of (a).

Figure 1.6: High-level structure of a parallelizing compiler.

Input to the compiler is usually a text representation of a source program. In practice, the source program can be a binary file, used in binary instrumentors and binary compilers [19, 174]; a *bytecode* file, such as is produced by a Java or Python source-to-byte code compiler, or a file with analysis information for the *compilation unit* included, to allow further analysis on a compilation unit.

A compilation unit is, as the name implies, the typical unit of a program over which the compiler operates, over which static symbol scopes are defined, and is the scope over which analysis and optimizations occur[1]. For C, C++ and Fortran, a compilation unit is typically a file. In the C and C++ languages the program has first passed through a preprocessor which inlines header files. These header files contain *declarations* and/or definitions of variables, types and functions – i.e., they give the name and type information about the object being declared, and may also define storage for the object. This allows better error checking by allowing the compiler to check for type mis-matches in assignments and function calls. For Java and some other languages, the compiler will check the declarations of functions in other files to see if they match what is expected. In Java programs, the unit of compilation for type checking is the entire program, whereas for optimization it is often a single function within a class. The javac Java-to-bytecode compiler performs no optimizations and the run time, or *dynamic*, compiler performs optimizations. Java optimizations are often restricted to

---

[1]Note, however, that many compilers also perform optimizations across compilation units, including optimizations across binary files. This is sometimes done by maintaining an intermediate representation (see Section 1.3.3) for multiple compilation units in the compiler This topic is beyond the scope of this lecture, however.

a single method or class to minimize the time spent compiling – an important consideration since the compiler is executing while the program is running.

1.3.2 PARSING

When a compilation unit is presented to the compiler, it is *lexically analyzed* and *parsed* by the compiler. Lexical analysis is the process of breaking up the string of characters that represent the source program into *tokens*, e.g., a variable or procedure name, a binary operator, a keyword, and so forth. Lexical analysis is a process that formally can be performed by a finite state machine. Parsing is the process by which the compiler recognizes different syntactic constructs formed from the tokens recognized by the lexical analyzer—a variable reference, a binary operation, an assignment statement, a function call, and so forth. Formally, parsing for most common languages can be performed by a slightly augmented push-down automata. Language designers often attempt to design languages that can be processed by these formal structures. Languages so designed can have lexical analyzers and parsers created for them automatically using tools such as Lex, YACC, Flex and Bison [143] and Antler [16]. Such languages also tend to be easy for humans to read.

The compiler will generally parse global declarations and definitions, which appear at the top of the file in C programs, and then compile, in turn, each function that appears in the program unit. The output of the parser is an *intermediate representation*, or IR, that represents the *semantics*, or meaning, of the program unit.

1.3.3 INTERMEDIATE REPRESENTATIONS

The IR of the program is repeatedly examined by different analysis passes, and modified to reflect the effect of optimizing transformations. The IR is represented as an *abstract syntax tree* [3, 17, 77] which is a graphical representation of the parsed program. In this work, we will modify this slightly and represent programs as a *control flow graph* (CFG), which is often imposed on the abstract syntax tree used by a compiler. A CFG is a graph whose nodes $b_i \in B$ are *basic blocks* and where an edge $b_i \rightarrow b_j$ means that block b_j may execute directly after b_i. A basic block is a sequence of operations with only one entry and one exit. Variations on the basic block are sometimes used in compilers, i.e., the basic block may contain several entry points (i.e., labels that can be branched to from outside the block) and one exit [87], or the basic block may only contain a single reference to a variable that can be accessed in a different thread [141]. The operations within a basic block may be represented as a sequence of low-level operations in three-address code, similar to what one might find in an assembly language program, or as expressions and statements represented as trees.

Figure 1.7(a) shows a fragment of a program, and Figure 1.7(b) shows the CFG for the program. Each block in the CFG is a basic block, and is labeled with the name of the block and the statements it contains. Figures 1.7(c) shows the *reverse* CFG for the program—each edge (a, b) is reversed to be an edge (b, a). The reverse control flow graph is used in the construction of the *static single assignment form* for programs (discussed shortly) in performing some kinds of *dataflow analysis* (discussed in Section 2.1), and for computing *control dependence* information (Section 2.4).

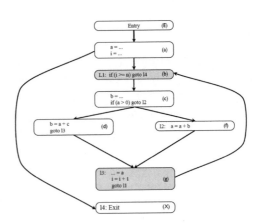

...

$S1 : a = b + c$
$S2 :$ for $(i = 0; i < n; i + +)\{$
$S3 :$ $b = \ldots$
$S4 :$ if $(a > 0)$
$S5 :$ $b = a + c;$
$S6 :$ else
$S7 :$ $a = a + b;$
$S8 : \}$

(a) A sample program.

(b) The control flow graph for the program of (a). Shaded nodes are in the *dominance frontier*.

$S1 :$ $a_0 = b_0 + c_0$
        $i_0 = 0$

$L1 :$ $i_1 = \phi(i_0, i_2)$
        $a_1 = \phi(a_0, a_2)$
$S3 :$ $b_1 = \ldots$
$S4 :$ if $(a_1 \leq 0)$ goto $\mathbf{L2} :$

$S5 :$ $b_1 = a_1 + c_0;$
$S5 :$ goto $\mathbf{L3} :$

$L2 : S7 : a_2 = a_1 + b_1;$

$L3 : S2 : i_2 = i_1 + 1$
$S2 :$ if $(i_2 < n)$ goto $\mathbf{L1} :$

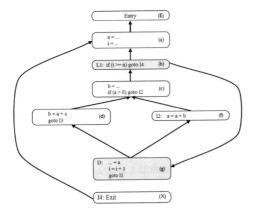

(c) The reverse control flow graph for the program of (a). Shaded nodes correspond to the *dominance frontier* in the forward graph.

(d) The program of (a) in SSA form.

Figure 1.7: Example of control flow and reverse control flow graphs, showing the dominator and post-dominator relationships.

	E	a	b	c	d	f	g	X
E								
a	✓	✓						
b	✓	✓	✓	✓	✓	✓	✓	
c				✓				
d					✓			
f						✓		
g				✓	✓	✓	✓	
X	✓	✓	✓	✓	✓	✓	✓	✓

(a) A table showing the dominates relationship for the reverse control flow graph of the program of Figure 1.7(a). Each row shows what nodes are dominated by the node at the start of the row, denoted with a ✓.

	E	a	b	c	d	f	g	X
E								
a			✓	✓	✓	✓	✓	✓
b				✓	✓	✓	✓	✓
c					✓	✓	✓	
d								
f								
g								
X								

(b) A table showing the dominates relationship for the reverse control flow graph of the program of Figure 1.7(a). Each row shows what nodes are dominated by the node at the start of the row, denoted with a ✓.

Figure 1.8: The dominator and post-dominator relationships for the program of Figure 1.7.

Static Single Assignment (SSA) form

The control flow graph and operations within the nodes of the basic blocks are sometimes converted to *static single assignment* (SSA) form [62, 75]. In the SSA form, storage is logically assigned a value once. We say "logically" because the single-assignment property is enforced during compilation and analysis, but the program is converted from the SSA form to a more traditional form for execution. Intuitively, the conversion into SSA form is done by giving the target of each assignment a unique name, and all uses of the value are changed to read from the unique name. Thus, in the program of Figure 1.7(c), the target of each assignment into a variable v is renamed v_j, v_{j+1}, and so forth. If a program consisted of only straight-line code, the each use of v would be replaced by a reference to v_j, where v_j is the name given the immediately preceding assignment to v. Often, arrays are treated similarly to scalars, that is a write (read) to an array is assumed to be to every element in the array. The Array SSA form of [120] overcomes this deficiency at the cost of some complexity in the compiler data structure and in runtime monitoring of stores to arrays when exact information about these cannot be gathered at compile time. Figures 1.7(d) shows the SSA form of the program of Figures 1.7(a).

Control flow complicates this, and must be handled. For example, in the program fragment of Figure 1.7(a), the value of a read in $S8$ can come from the assignment of a in either $S1$ or $S7$. To deal with this, the SSA form uses ϕ functions to choose between the different values of a that arrive at a program point, and assign the value to a new, uniquely numbered a.

The algorithm for placing *Phi*-functions uses the concept of the *dominance frontier* of a statement containing an assignment to a. A basic block b_j is *dominated* by a basic block b_d if all paths from the start of the program to b_j must pass through b_d. Figure 1.8(a) and (b) shows the dominance relationships for the CFG and reverse CFGs, respectively, for the program of Figures 1.7(a). The dominance frontier of some statement b_j is all basic blocks b_{df} such that b_{df} dominates a predecessor of b_j in the CFG but does not dominate b_j [62]. More intuitively, the dominance frontier of b_j contains the blocks where paths in the control flow graph that bypass b_j join with the path that contains b_j. One way of finding the dominance frontier for a CFG is to compute the *control dependence nodes* in the reverse CFG using the concept of dominators and post-dominators in the reverse CFG—these nodes are the dominance frontier. Intuitively a node X is control dependent on a node Y if the execution of Y controls whether or not X is executed. The concepts of control dependence and dominance frontiers are discussed in Section 2.4.

The SSA form is useful because it allows a variety of scalar optimizations such as constant propagation [240] (which finds variables that hold constant values) and *use-def* computation to be done efficiently. Most modern compilers have an IR that is in SSA form.

Other compiler structures in the intermediate representation

The IR of a program maintains other useful information about the program. The two most important data structures not yet discussed are the *symbol table* and *call graph*.

The symbol table contains information about every symbol in the program that can be accessed from the function or compilation unit being compiled. This includes classes, variables, functions, structures, and global and external variables and functions. Standard information maintained consists of variable types, return types and parameter types and number for functions, structure fields and their types, global and external variable types, whether variables are constants, and member functions of variables and class variables. Information such as whether a variable is constant may be derived by the parser when the variable declaration is parsed (if it is declared as "const"), or may be discovered later as a result of analysis.

The call graph (CG) has a node for each unique function or method in the program (or compilation unit). An edge $n_i \rightarrow n_j$ exists between two nodes in the CG if function n_i calls function n_j. If there is a path from n_i to itself through zero or more other nodes, then n_i is recursively invoked. The nodes of the CG often contain a CFG that represents the actions of the function, as described above.

Compiler optimizations for sequential and parallel machines

Compilers perform several different flavors of analysis and optimization. In general, the goals are to reduce the amount of work performed in a computation, reduce the amount of storage used, and to enable the work that must be performed to be performed in parallel. Examples of optimizations that reduce the amount of work performed include, but are not limited to:

dead code elimination which finds never executed code and removes it from the program. Dead code often results from flags used to support debugging (e.g., if (debug > 0) ...) or as a side affect of other compiler transformations;

function inlining which replaces a call to a function by the body of the function, i.e., the function is treated similarly to a macro. In the case of small *accessor* functions in object oriented programs, the benefits of not having to go through a full function invocation to access a class field can be substantial;

tiling which seeks to reduce the overhead of cache misses;

strength reduction which replaces an operation (e.g., integer division by a power of two) with a cheaper operation (e.g., a right shift.) We note that even something as simple as this can become complicated. For example, some languages like Java round towards zero, whereas doing a right shift on a negative complex numbers will result in a rounding down. Thus, $-5/2$ should give a result of -2 in Java, but shifting right yields ($1101 >> 1 \rightarrow 1101$, or -3. Handling such subtleties manually is error prone and a good reason to rely on compiler optimizations, especially for low-level optimizations.

An example of a transformation that enables sequential optimizations is function inlining. By bringing the called (or *callee*) function code inside of the calling (or *caller*) function, *intra*-procedural analysis techniques can be used to analyze and optimize the code in the called function along with the results in the callee. By creating a new copy of the function at each site where it is called and inlined, optimizations that are legal, or beneficial, only for the invocations at a single call site are enabled.

At a much lower level, *register allocation* is used to maximize the number of variable reads and writes that are into and out of registers while minimizing the number of loads and stores of values between registers and memory. Also at the low level, instruction scheduling can be used to order operations so that the hardware is likely to execute them in a way that maximizes instruction level parallelism.

The details of most of these optimizations is beyond the scope of this lecture. In Chapter 8, pointers to papers giving a good overview of these techniques can be found. The bulk of this lecture will discuss compiler support for parallel machines, as described in the next section. Nevertheless, parallel programs consist of collections of sequential code executing in parallel, and compiler techniques that optimize this sequential code also improve the performance of the program.

1.4   COMPILER SUPPORT FOR PARALLEL MACHINES

Ideally, a compiler would take a sequential program and automatically translate it into a parallel program that efficiently makes use of the cores in a multicore processor. With current technologies, a loop can execute in parallel if either (i) each iteration of the loop executes independently of all other iterations, or (ii) properties such as commutativity can be exploited to allow different iterations write and read the same data to execute in parallel. Figure 1.9 shows a program whose loops can be parallelized by transforming its loops into ones that meets the above criteria.

Consider first the i loop. Each iteration of the loop writes a unique memory location b[i], and no iteration of the loop reads this location. Thus, no iteration of the loop reads a memory location

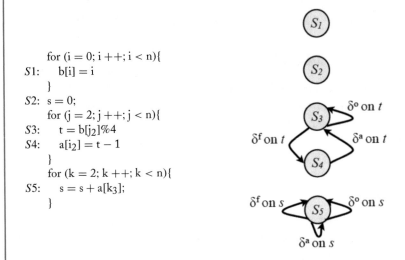

```
        for (i = 0; i ++; i < n){
S1:     b[i] = i
        }
S2:  s = 0;
        for (j = 2; j ++; j < n){
S3:     t = b[j₂]%4
S4:     a[i₂] = t − 1
        }
        for (k = 2; k ++; k < n){
S5:     s = s + a[k₃];
        }
```

(a) A program.   (b) A dependence graph for the program.

```
        parfor (i = 0; i ++; i < n)
S1:     b[i] = i
S2:  s = 0;
        int numThreads = getThreadCount();
        int myThreadID = whoAmI();
        for (j = 2; j ++; j < n) {
            int t′[n];
            int s′[n];
S3:     t′[i] = b[j]%4
S4:     a[j] = t′[i] − 1
        }
        for (k = 0; k < numThreads; k ++)
S5:     S[myThreadID] = S[myThreadID] + a[k];
        for (k′ = 0; k′ < numThreads; k′ ++)
S6:     s+ = S[k′];
```

(c) The program of (a) transformed and parallelized.

Figure 1.9: An example of a loop with multiple dependences.

touched by another loop, every iteration of the loop is independent of all others, and the loop can be easily run in parallel. For the j loop the situation is more complicated. Every iteration of the loop reads and writes t. Should the loop, as is, be run in parallel, there will be many iterations trying to read and write the same memory location, and an execution of the program will probably not give a correct result. Finally, in the k loop every iteration also reads and writes the variable s.

Compilers use *dependence analysis*, described in Chapter 2, to determine if accesses to an array may be to the same element of the array, and if so, how many iterations separate those accesses. In Figure 1.9(b) the *dependence graph* for the program is shown. The dependence graph summarizes the problematic accesses of memory on t and s described above. δ^f is a *flow* dependence, which indicates a location is written, and then later read later on, δ^o is an *output* dependence, which indicates a the same location is written, and then later written again one or more times, and δ^a is an *anti* dependence, which indicates a location is read, and then later written. We note that flow, anti and output dependences correspond to read-after-write (RAW), write-after-read (WAR) and write-after-write (WAW) hazards, respectively. The concepts of dependence will be refined and discussed in greater detail in Chapter 2.

In the loop of the example of Figure 1.9, the accesses to t can be made independent by replacing the scalar t with an array t' that has one element for each iteration of the loop, and replacing each reference to t with a reference to $t'[j]^2$. Doing this leads to the code in Figure 1.9(c). This, and other transformations to eliminate or modify dependences to enable parallelization, are described in Chapter 4.

The behavior of s in the k loop looks similar to that of t, but there is an important difference. Because the only operations on s in the loop are commutative, the final value of s can be computed as a *reduction*. As with t, s is expanded into an array s' with one element for each thread. To get the final value of s, however, it is necessary to sum up the numThreads elements of s' into s, as is done by the sequential k' loop. Chapter 5 discusses reduction recognition, and how code is generated for reductions.

In Chapter 3, we discuss how the analyses used to parallelize programs can be used to perform other optimizations. The most widely used of these is *tiling*, which alters the order of array accesses in order to increase cache locality, but the analyses can support a variety of other transformations.

The analyses and optimizations discussed so far, and through Chapter 3 assume the input program is sequential. If it is not, then issues related to those raised by implementing cache coherence protocols on shared memory processors are raised. In Section 2.6 we discuss how compilers can deal with these issues arising from compiling programs that are already parallel.

An overview of compiling for distributed memory machines is given in Chapter 6. Whereas communication across threads executing in parallel on a shared memory machine is done via loads and stores, communication on a distributed memory machines requires explicitly calls to communication library code, and carries a significantly higher startup cost than a load or a store. This, in turn, makes

---

[2]A more efficient allocation of t' can be made by realizing that only one element is needed per thread, as discussed in Section 4.8.

it necessary to minimize the number of communication operations that occur, and makes it essential that each operation deliver as much data as possible, placing a unique burden on the compiler.

In Chapter 7, we discuss the solution of Diophantine equations, which are equations that whose solutions and coefficients are integers. Diophantine equations are used in several places by compilers.

Finally, in Chapter 8, we present a guide for further reading both on the topics discussed in this work, and in related and more advanced topics that were not covered.

CHAPTER 2

Dependence analysis, dependence graphs and alias analysis

As described in Chapter 1, determining what data (memory) is accessed by references in the program is fundamental to determining if different regions of a program can execute in parallel. Three major techniques are used to determine when two references access the same storage: *dependence*, *use-def* and *alias* analysis. In general, dependence analysis is used to analyze array accesses, and alias and use-def analysis are used to analyze accesses to scalars or data that is treated as a scalar.

All of the analyses we discuss have the property that they are *sound* or *conservative*. Any analysis performed at compile time will be forced to approximate the behavior of the program at run time, and this approximation will cause the results of the analysis to differ from what actually happens at run time. The question is, what effect do these differences have on the correctness of transformations using the analysis results? If the analysis is sound, or conservative, then the differences will cause optimization opportunities to be missed, but will not lead to incorrect executions of the program. What constitutes a conservative result from an analysis is in part a function of how the results are used. Consider an analysis that determines if two reads are to the same memory location. If the analysis is used by a pass that allows the results of the first read, a_1, stored in a register to be reused by a second access, a_2, then the analysis is conservative only if every pair of accesses a_1, a_2 said to be to the same location *must* be to the same location. If they are not, then a_1 will read the wrong value. In this case, the analysis is conservative if it only says that accesses are to the same location if they always are. Now consider when the analysis is being used to determine if a read of storage by a_2 can be moved above a write by a_1. In this case, if the analysis says that two accesses that may, or may not, access the same storage do not access the same storage (a conservative result for the first use of the analysis) then illegal moves of two accesses past each other may occur. In this case, the analysis should say that two accesses are to the same location if they *may* be to the same location. It is entirely the responsibility of the creator of an analysis and the user of the analysis (the client) to ensure that the approximations made by the analysis are appropriately conservative.

We will first describe a basic *dataflow* analysis, and use the concepts developed to discuss the more complicated alias analysis. We will then move on to dependence analysis and conclude with a discussion of use-def analyses. Dataflow analysis has been a area of intense research for over 40 years ([100, 200]), and a more extensive introduction can be found in [3, 17, 77].

2.1   DATAFLOW ANALYSIS

We will first discuss a simple dataflow analysis to allow a clearer explanation of certain concepts to be done. The analysis we will consider is *constant propagation*, which attempts to determine what values in a program are constants.

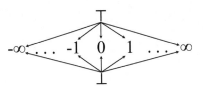

$\wedge$	$\top$	c_1	c_2	$\bot$
$\top$	$\top$	c_1	c_2	$\bot$
c_1	c_1	c_1	$\bot$	$\bot$
c_2	c_2	$\bot$	c_2	$\bot$
$\bot$	$\bot$	$\bot$	$\bot$	$\bot$

(a) the semi-lattice used by constant propagation.

(b) The join operation $\wedge$ for the semi-lattice.

Figure 2.1: The semi-lattice used in the constant propagation analysis.

2.1.1   CONSTANT PROPAGATION

Constant propagation, like most dataflow analyses, is performed by traversing the control flow and calling graph of the program. We will first discuss the simpler intra-procedural case (i.e., where the analysis is only examining a single procedure) where only the control flow graph of the function being analyzed is needed, using the program of Figure 2.2(a) and the control flow graph of Figure 2.2(b). Prior to performing the analysis, each node in the graph is annotated with two pieces of information: facts that become true in the analysis when the node is visited, and facts that are no longer true after the node is visited. The specification of facts that become true is contained in the *GEN* set, and the specification of facts that are no longer true after the node is visited are given by the *KILL* set. For constant propagation, we can say three things about the domain of values for variables in our analysis:

1. it is a constant with a value v;

2. we have not yet discovered anything about the variable; and

3. we have discovered something about the variable, and it is *not* a constant.

The relationship between these values is often represented by a *semi-lattice L*, which consists of a set of values V and a *meet* operator $\wedge$.

   The values in the semi-lattice and the meet operator vary from analysis to analysis. In Figure 2.1(a) a diagram showing the values of the semi-lattice for a constant propagation is given. Given two values v_i and v_j, let v_r be the first common descendant of v_i and v_j in the semi-lattice, then $v_r = v_i \wedge v_j$. Using these definitions, we see how constant propagation can be done using the semi-lattice shown in the figure.

```
      void foo(int k) {
S1      int j;
S2      k = 4;
S3      read(j);
S4      for (i = 1; i < n; i ++) {
S5        if (j > 0) {
S6          k = 5;
S7          j = k;
S8        }
S9      }
      }
```

(a) A program over which constant propagation will be performed.

S1 IN = ∅, GEN = ∅, OUT = {[i,⊤], [j,⊤], [k,⊤]}
S2 IN = ∅, GEN = {[k,4]}, OUT = {[i,⊤], [j,⊤], [k,⊤]}
S3 IN = ∅, GEN = {[j,⊥]}, OUT = {[i,⊤], [j,⊤], [k,⊤]}
S4' IN = ∅, GEN = {[i,0]}, OUT = {[i,⊤], [j,⊤], [k,⊤]}
S4 IN = ∅, GEN = ∅, OUT = {[i,⊤], [j,⊤], [k,⊤]}
S5 IN = ∅, GEN = ∅, OUT = {[i,⊤], [j,⊤], [k,⊤]}
S6 IN = ∅ GEN = {[k,5]}, OUT = {[i,⊤], [j,⊤], [k,⊤]}
S7 IN = ∅ GEN = {[j,k]}, OUT = {[i,⊤], [j,⊤], [k,⊤]}
S8 IN = ∅, GEN = {[i,i+1]}, OUT = {[i,⊤], [j,⊤], [k,⊤]}
S9 IN =∅, GEN = ∅,OUT = {[i,⊤], [j,⊤], [k,⊤]}

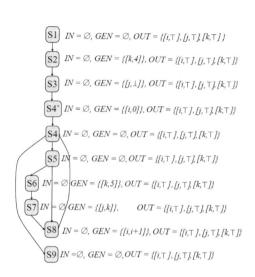

(b) The CFG for the program of (a) initialized to begin the constant propagation dataflow analysis.

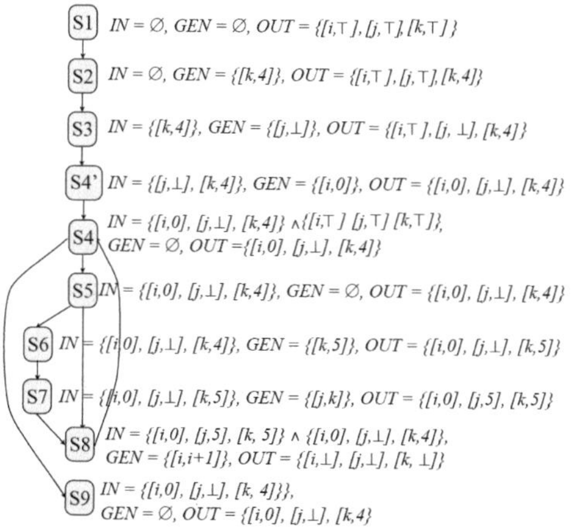

S1 IN = ∅, GEN = ∅, OUT = {[i,⊤], [j,⊤], [k,⊤]}
S2 IN = ∅, GEN = {[k,4]}, OUT = {[i,⊤], [j,⊤], [k,4]}
S3 IN = {[k,4]}, GEN = {[j,⊥]}, OUT = {[i,⊤], [j, ⊥], [k,4]}
S4' IN = {[j,⊥], [k,4]}, GEN = {[i,0]}, OUT = {[i,0], [j,⊥], [k,4]}
S4 IN = {[i,0], [j,⊥], [k,4]} ∧{[i,⊤] [j,⊤] [k,⊤]}, GEN = ∅, OUT ={[i,0], [j,⊥], [k,4]}
S5 IN = {[i,0], [j,⊥], [k,4]}, GEN = ∅, OUT = {[i,0], [j,⊥], [k,4]}
S6 IN = {[i,0], [j,⊥], [k,4]}, GEN = {[k,5]}, OUT = {[i,0], [j,⊥], [k,5]}
S7 IN = {[i,0], [j,⊥], [k,5]}, GEN = {[j,k]}, OUT = {[i,0], [j,5], [k,5]}
S8 IN = {[i,0], [j,5], [k, 5]} ∧ {[i,0], [j,⊥], [k,4]}, GEN = {[i,i+1]}, OUT = {[i,⊥], [j,⊥], [k, ⊥]}
S9 IN = {[i,0], [j,⊥], [k, 4]}}, GEN = ∅, OUT = {[i,0], [j,⊥], [k,4}

(c) The CFG after one pass.

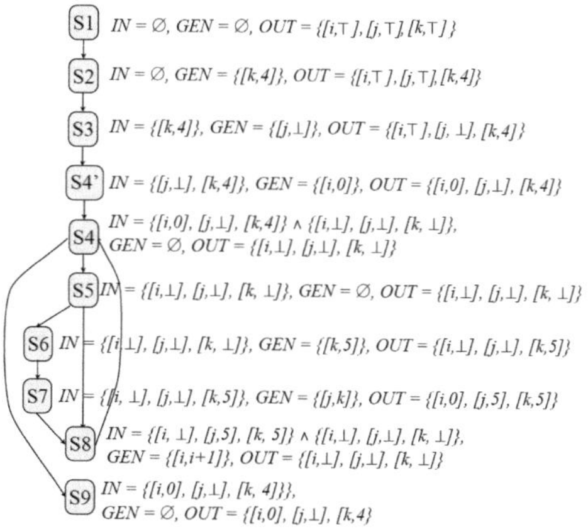

S1 IN = ∅, GEN = ∅, OUT = {[i,⊤], [j,⊤], [k,⊤]}
S2 IN = ∅, GEN = {[k,4]}, OUT = {[i,⊤], [j,⊤], [k,4]}
S3 IN = {[k,4]}, GEN = {[j,⊥]}, OUT = {[i,⊤], [j, ⊥], [k,4]}
S4' IN = {[j,⊥], [k,4]}, GEN = {[i,0]}, OUT = {[i,0], [j,⊥], [k,4]}
S4 IN = {[i,0], [j,⊥], [k,4]} ∧ {[i,⊥], [j,⊥], [k, ⊥]}, GEN = ∅, OUT = {[i,⊥], [j,⊥], [k, ⊥]}
S5 IN = {[i,⊥], [j,⊥], [k, ⊥]}, GEN = ∅, OUT = {[i,⊥], [j,⊥], [k, ⊥]}
S6 IN = {[i,⊥], [j,⊥], [k, ⊥]}, GEN = {[k,5]}, OUT = {[i,⊥], [j,⊥], [k,5]}
S7 IN = {[i, ⊥], [j,⊥], [k,5]}, GEN = {[j,k]}, OUT = {[i,0], [j,5], [k,5]}
S8 IN = {[i, ⊥], [j,5], [k, 5]} ∧ {[i,⊥], [j,⊥], [k, ⊥]}, GEN = {[i,i+1]}, OUT = {[i,⊥], [j,⊥], [k, ⊥]}
S9 IN = {[i,0], [j,⊥], [k, 4]}}, GEN = ∅, OUT = {[i,0], [j,⊥], [k,4}

(d) The CFG after the second pass—the dataflow information has converged.

Figure 2.2: An example of a flow sensitive constant propagation analysis.

Constant propagation is a *forward* data flow analysis, that is, facts that are true at some point are propagated forward through the program. The program state, before a statement is executed, is in the *IN* set, and the program state after the statement is executed is in the *OUT* set. Moreover, at each node in the control flow graph certain facts become true—these are represented by a *GEN* set associated with each node of the program. Also at each node in the CFG certain facts must become false—these are represented by a *KILL* set associated with the node.

A transfer function of the form $OUT = GEN \wedge (IN - KILL)$ represents the effect of the node on the program state when the statement is executed. Almost all dataflow analyses use a transfer function of this general form. The set associated with the nodes must first be initialized. In constant propagation, the *KILL* set is empty, and is set as such for every node. The *IN*, *OUT* and *GEN* sets consist of doubles of the form [v, *val*]. For the *IN* and *OUT* sets, *val* is an element of the semi-lattice. For the *GEN* set, v is a variable that is on the left-hand side of the assignment, and *val* is either a single variable, a constant, or an expression of the form $val_i \otimes val_j$ that is assigned into that value. Figure 2.2(b) shows the nodes of the CFG annotated with the *GEN* and (empty) *KILL* sets that capture the effect on the nodes on the program state, and the initial values of the *IN* and *OUT* sets.

The transfer function is evaluated as follows. For simplicity, we assume one statement per node, and each node contains one of five kinds of expressions:

$$
\begin{aligned}
v_i &= c_i \\
v_i &= v_j \\
v_i &= op_1 \otimes op_1 \\
v_i &= foo(\ldots), \\
null &
\end{aligned}
$$

where c_i is a constant literal, v_i, v_j are variables, op_1 and op_2 are either variables or constants, *foo* is a function and *null* is a node where no variables are assigned. Let $VAL_S(t) = val$, where t is the double [v, *val*] in the set S, i.e., *IN*, *GEN* or *KILL*. Let $VAR_S(t) = v$, where t is the same double. For every double [v_i, *val*] in the *IN* set with no corresponding double [v_i, ...] in the *GEN* set, the double [v_i, *val*] is simply copied to the *OUT* set. This is true for all variables when the expression is *null*. For the variable v_i such that [v_i, *val*] is in the *GEN* set, the following action is taken. If *val* is a variable v_j, then the double [v_j, *val*] is found in the *IN* set (i.e., the current value of v_j in the domain represented by the semi-lattice is found), and the double [v_i, v_j] is placed into the *OUT* set. If *val* is the constant c_j, then the double [v_i, c_j] is added to the *OUT* set. Otherwise, the expression in the node is $val = op_1 \otimes op_2$. As was just done with a singleton right-hand side, we look up each of op_1 and op_2 in the *IN* set. If both are constants or constant valued variables, the expression $op_1 \otimes op_2$ is evaluated using the constant values. The double [v_i, r] is then added to the *OUT* set, where r is the result of evaluating the expression. Finally, if either op_1 or op_2 is not a constant, the double [v_i, $VAL(op_1) \wedge VAL(op_2)$] is added to the *OUT* set. In Figure 2.2(c), the result of doing this throughout the program on a single pass is shown.

Statements like the C `for` statement may be broken into more than one statement—this has been done for *S4*. Statement *S4* is broken into two statements, one which assigns 0 to `i`, and the other which performs the increment of `i`.

Statement *S4* is a *join* point in the control flow graph. As a result, there are values for each variable v_i defined along each control path entering the join. At this particular join point, the values along the path from the top of the program will hold the first time statement *S4* is visited, and values from within the loop will hold the second and later times that it is visited. The analysis, however, must represent values that hold at any execution of *S4*. To model this, the *IN* set of *S4* contains the meet of the two *OUT* sets, that is, for every double $t_{S4'}$ (t_{S8}), such that $VAR_{OUT}(t_{S4'}) = VAR_{OUT}(t_{S8})$, the double $[VAR_{OUT}(t_{S4'}), VAL_{OUT}(t_{S4'}) \wedge VAL_{OUT}(t_{S8})]$ is in the *IN* set of *S4*.

The propagation of values through the CFG continues until no changes occur, and the algorithm terminates with the graph as it appears in Figure 2.2(d). The contents of the *IN* set associated with each statement tells whether the corresponding variables are constants or not, and if so, what their value is, immediately before the statement executes. We know the algorithm will terminate because the result of the $\wedge$ operation either moves down the semi-lattice, or stays in the same place. Thus, at any statement *S* there can only be *h* updates of the variable's value in *S*, where *h* is the height of the semi-lattice.

The analysis just described is flow sensitive, that is, there is an *IN* set for each statement, and the *IN* set describes the state of the execution along the program flow in the CFG for each statement, and as such, all facts in the *OUT* set are conservatively correct for each point in the program. A *flow insensitive* analysis can be performed where there is effectively a single *IN* set that also serves as the *OUT* set for every statement. As a statement is visited, the *IN* set is updated, and then used as the input for the analysis of the next statement. Flow insensitive analyses are generally less precise, and more conservative, than flow sensitive analyses because a fact is true only if it is true throughout the entire control flow graph. Thus, the knowledge in the flow sensitive algorithm that k = 5 in *S6*, and j = k = 5 in *S7*, but have different constant values in other parts of the program, is lost in the flow insensitive analysis. Flow insensitive algorithms tend, however, to be faster than flow sensitive algorithms. Figure 2.3 shows the effect of flow insensitive analysis on the example of Figure 2.2.

What happens when a procedure call is encountered? When the intra-procedural analysis of a function is finished, its effect on the global state of the program can be represented as a *summary*. Thus, in the example of Figure 2.2, the parameter *k* is not a constant upon leaving the function. Thus, at each call site for `foo`, the *GEN* set for the calling statement would contain the double $\{[\kappa, \bot]\}$ to reflect this, where κ is the argument passed to the parameter k of `foo`. Similarly, the state of the program at the point of the call can be represented as the *IN* set of the function call statement. The analysis, upon encountering the function call, would send the values associated with arguments to the intra-procedural analysis of the function body. Analogous to flow sensitive and insensitive analyses, inter-procedural analyses can be *context sensitive* and *context insensitive*. In a context sensitive analysis, each function call site has a separate summary (*GEN* and *KILL*) sets, whereas in a context

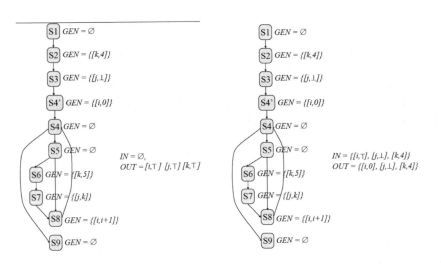

(a) The initial flow insensitive structures for the program of Figure 2.2(a).

(b) The flow insensitive structures after visiting $S4'$.

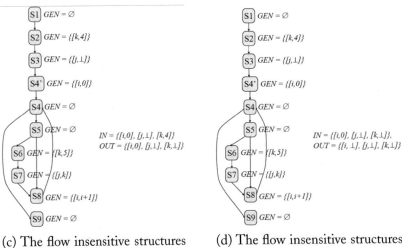

(c) The flow insensitive structures after visiting $S7$.

(d) The flow insensitive structures after visiting $S8$ and converging.

Figure 2.3: A flow insensitive constant propagation analysis on the program of Figure 2.2(a).

insensitive analysis all call sites share summary *IN* information. Similar precision/speed trade-offs exist as with the flow sensitive and insensitive analyses.

A work queue is typically used to implement a dataflow algorithm. The first node of the function is placed into the queue. Until the queue is empty, the algorithm removes a node from the queue. The *IN* set is computed from the predecessor node's *OUT* sets, and the transfer function is evaluated. If the *OUT* set of the node being processed changes, its successor nodes are placed into the work queue. This causes nodes to be reevaluated only when their inputs change.

Finally, some analyses, such as live variable analysis, are backwards analyses, and information is propagated backwards through the program. In live variable analysis, the fact that a variable is read later in a program makes a value in the variable at an earlier point in the program *live*. Backwards analyses are in principle just like forward analyses, except that a reverse control flow graph is used. A reverse control flow graph is formed from the control flow graph by changing each edge $\alpha \rightarrow \beta$ to $\beta \rightarrow \alpha$. In these analyses the *OUT* sets are computed from the *IN* sets of the predecessor nodes in the *reverse* control flow graph, and the transfer function then computes the *IN* nodes value.

2.1.2   ALIAS ANALYSIS

Alias analysis is a dataflow analysis that tries to determine the different "names" by which an object in memory is referenced. By "object" we do not mean an object in the object oriented use of the term, but rather any collection of memory that can be referenced in whole by the program. An object in our sense may be a word, a structure, the field of a structure, an entire array or some other element (including an object in the OO sense of the word) in the program. Two different pointers can be made to refer to the same object by assigning the same value, or address, to each. Two different function parameters can be made to refer to the same object in languages that pass arguments to functions by reference (such as Fortran) by passing the same object as two or more arguments. In languages such as C and C++ that pass arguments by value, passing two or more pointers that contain the same address as parameters causes aliasing. Address arithmetic can also cause real, or apparent aliasing.

We will refer to the program of Figure 2.4(a) when describing alias analysis. In Figure 2.4(b) the control flow graph (as described in Section 1.3.3) for the program is shown.

The program first declares three pointers, p, q and r, in statements $S1$, $S2$ and $S3$, and assigns to each pointer the address of a heap allocated object. This raises an interesting problem in alias analysis—how does the analysis refer to a heap allocated object? Note that if the allocation site is executed repeatedly, as in:

$$\text{for } (i = 0; i < n; i {+}{+}) \{$$

$$\cdots$$

$$S_j \qquad \text{ptr} = \text{malloc}(\ldots)$$

$$\cdots$$

$$\}$$

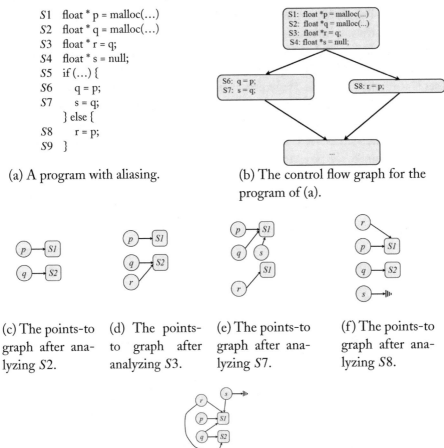

$S1$ float * p = malloc(...)
$S2$ float * q = malloc(...)
$S3$ float * r = q;
$S4$ float * s = null;
$S5$ if (...) {
$S6$ q = p;
$S7$ s = q;
      } else {
$S8$ r = p;
$S9$ }

(a) A program with aliasing.

(b) The control flow graph for the program of (a).

(c) The points-to graph after analyzing $S2$.

(d) The points-to graph after analyzing $S3$.

(e) The points-to graph after analyzing $S7$.

(f) The points-to graph after analyzing $S8$.

(g) The points-to graph after analyzing $S9$.

Figure 2.4: Alias analysis example.

then the variable `ptr` may reference a number of different objects, the number of which is unknown at compile time. Therefore, the different objects cannot all be uniquely and individually represented at compile time, and that which cannot be represented by a compiler cannot be analyzed by the compiler. One standard way around this representation problem is to name allocation sites, and refer to all objects allocated at that site by the site name. Thus, all objects allocated in statement S_j in the loop above would be referred to as an S_j object, and any action taken on any of those objects would be assumed to take place on all of them.

Returning to the example of Figure 2.4, in statements $S1$ and $S2$ the assignments to p and q cause them to point to objects $S1$ and $S2$, respectively. This is represented by the *points-to* graph shown in Figure 2.4(c). In the simple points-to graph of this example, each node is a circle labeled with the name of a pointer, or a square labeled with the name of an allocation site representing all of the objects allocated at the site. Edges in the graph go from a pointer to the object referenced by the pointer. In $S3$, r receives the value of q, and therefore points-to object $S2$, as shown in Figure 2.4(d). Because q and r point to the same object, they are aliased at this point in the program.

The next statement is an if statement. At this point we will assume a *flow sensitive* analysis, which not only follows the control flow of the program but keeps analysis information at each point in the program. Along the true path of the if the statement q = p is executed. This makes q point-to what p points-to instead of pointing to object $S2$. As well, the assignment s = q makes s point to what q (and p) point to. This is shown in the points-to graph of Figure 2.4(e). Along the false path r = p is executed, and r now points-to object $S1$ rather than object $S2$, as shown in Figure 2.4(f). At the end of the if statement, the points-to graph must be conservative and correct regardless of which path the program took. Thus, whatever is true along either path reaching this point must be true at $S9$. Since p points only to $S1$ along both paths, it must point to $S1$ at the join of the two paths. Because q points-to $S1$ at the end of the true branch, and to $S2$ at the end of the false branch, it can point to either at the end of the if. r points to $S1$ or $S2$ at the end of the if, and s points-to either $S1$ or nothing. This is shown in Figure 2.4(g).

In this analysis the result of the join operation for a graph $p \rightarrow q$ and $p \rightarrow r$ is a graph where p is pointing to both q and r, i.e., is the union of the graphs. The domain of the semi-lattice over which this analysis operates contains increasing connected points-to graphs containing all heap objects and pointers visible in the region of the program being analyzed. Like the infinitely wide semi-lattice using in constant propagation, this semi-lattice is never explicitly represented.

The analysis can be *flow insensitive* instead of flow sensitive. In the flow insensitive analysis, a single points-to graph is maintained for the entire program region under analysis, rather than at each program point, and the resulting points-to graph, when the analysis is finished, must be conservatively true at all points in the region under analysis. More concretely, in our example, when $S9$ is analyzed, q is made to point to both $S1$ and $S2$, because it points-to each at some location in the program. Similarly, when s = q is analyzed, s must point to both $S1$ and $S2$ because q points-to both of these. At the merge point at the end of the if no additional action is necessary because the merger of information has already happened on the single points-to graph used by the flow insensitive analysis.

Alias analysis is imprecise for several reasons. Context and flow insensitive analyses merge information from multiple points in the program, and consider the information to hold at all points in the program. These less precise analyses are used because they are generally faster than the more precise context and flow sensitive analyses. Because the analysis must be true for all paths through the program, merging alias information at control flow join points, such as after the if in the example above, also loses information about what is true during a particular execution of the program at this point. Finally, the fact that all objects allocated at a particular site are treated as the same object, and that arrays are generally treated as a single, monolithic object, such that a write (read) to any element of the array is considered a write (read) to the entire array, also causes alias analyses to be generally conservative. In Section 2.3 we discuss *data dependence* analysis, which aims to gather more precise information about array accesses.

2.2   ABSTRACT INTERPRETATION

Abstract interpretation [58, 59] is an alternative approach to dataflow analysis for the static analysis of computer programs. Abstract interpretation has traditionally been less widely used than dataflow analysis in compilers, but is an area of intense research interest, and arguably is gaining in popularity.

Conceptually, a programming language $\mathbf{L}$ has some semantics $\mathcal{S}$ and operates over some domain $\mathcal{D}$. Thus, the semantics of the programming language specifies how programs map input values (in the domain $\mathcal{D}$) onto the domain $\mathcal{D}$. Given a program $\mathbf{p} \in \mathbf{L}$, the semantics $\mathcal{S}$ define allowable outcomes in $\mathcal{D}$ given some input or initial state. Moreover, this outcome can be computed by executing the sub-steps of the program in a way allowed by the language semantics.

We note that programs typically have an infinite number of outcomes, and computing all of these outcomes at compile time is clearly impossible. Abstract interpretation, like dataflow analysis, deals with this by defining a restricted, or *abstract*, domain for the analysis to operate on. In our dataflow analysis constant propagation example above, values of variables could be either $\top$, $\bot$, or a constant value. An abstract interpretation designed for performing constant propagation might have as its abstract domain $\mathcal{D}^{\#}$ the same elements, i.e., $\mathcal{D}^{\#} = \{\top, \bot, i \in \mathbb{Q}\}$. Thus, in the abstract domain the program would perform a mapping onto $\mathcal{D}^{\#}$ instead of $\mathcal{D}$.

It is incumbent on the designer of the abstract interpreter to provide a mapping M from $\mathcal{D} \mapsto \mathcal{D}^{\#}$, and to prove certain properties of this mapping. For example, it is typically shown that given a binary operation $\oplus(v_1, v_1) \mapsto v_3, v_1, v_2, v_3 \in \mathcal{D}$, that the corresponding binary operation operating on the abstract domain gives a consistent result, i.e., $\oplus(v_1^{\#}, v_1^{\#}) \mapsto v_3^{\#}, v_1^{\#}, v_2^{\#}, v_3^{\#} \in \mathcal{D}^{\#}$, and if $M(v_1) \mapsto v_1^{\#}$, and if $M(v_2) \mapsto v_2^{\#}$, then it must be that $M(v_3) \mapsto v_3^{\#}$. Composition, associativity, commutativity, and other properties defined by the semantics $\mathcal{S}$ on $\mathcal{D}$ should also hold for the operations on $\mathcal{D}^{\#}$

To perform the analysis, the statements of the program are interpreted over this abstract domain, and by showing that operations over the program state in $\mathcal{D}^{\#}$ lead to monotonically increasing values of the program state, it can be shown that the interpretation will reach a fixed point such that further interpretation does not change the program state. This final program state then is the result of the analysis.

A comparison of dataflow analyses and abstract interpretation   Dataflow analysis and abstract interpretation have their relative strength and weaknesses. In practical terms, many commercial compilers (and open source compilers for Java, C and C++) contain built-in dataflow analysis frameworks that allow the analysis writer to only define the transfer functions and computation of IN (OUT) sets from the predecessors (successors) for forward (backward) dataflow analyses. This removes much of the complexity of implementing many dataflow analyses. Some analyses, such as liveness, which determines if a value in a variable may be used later in the program execution, are more difficult to do with abstract interpretation.

At the same time, abstract interpretation makes the linkage between the semantics of the language being analyzed and the analysis semantics explicit. In particular, each value in the abstract semantics corresponds to a well defined set of values that the actual program can take on. In dataflow analyses these relationships must be independently proven. Nevertheless, it is often felt that dataflow analyses are more intuitive, and consequently are more likely to be taught in introductory compiler courses. It has been shown [208, 221, 222] that dataflow analysis is equivalent to model checking using mu-modal calculus. Schmidt [208] makes the point that

> classic [data flow analyses] merely scratch the surface as to the forms of iterative static analysis one can perform – the combination of the abstract interpretation design space and the modal mu-calculus define the design space of iteratively static analyses.

2.3   DATA DEPENDENCE ANALYSIS

Consider the loop of Figure 2.5(a). A standard alias analysis would consider the b array to be a monolithic structure, and would conservatively consider the two accesses of b to touch the same memory location. A closer examination of the accesses, however, shows that b[2*i] is only writing the even elements of the array, and b[2*i-1] is only reading the odd elements of the array, and the same location is never touched twice. The goal of dependence analysis is to provide a more precise analysis than alias analysis that can detect when this happens. Moreover, when the same array element is accessed two or more times, dependence analysis will try and identify the order that the accesses may occur, and if possible to determine how many iterations of the loop nest there are between the two accesses. Because dependence analysis is conservative, it seeks to find those cases when it can be proven that a dependence does not exist between two references, and otherwise assumes that a dependence exists. Depending on the particular dependence analysis technique, it can sometimes prove the harder claim that a dependence really exists, but in general dependence analysis seeks to show when a dependence cannot exist rather than when it must.

There are four kinds of data dependence, three of which are commonly considered in loop parallelization. A *flow* or *true* dependence exists when a value is written to an array element, and then read from that element at a later time. An *anti* dependence exists when a value is read from an array element, and then written to that element at a later time. An *output* dependence exists when a value is written to an array element, and then overwritten with a different value at a later time. The fourth type of dependence is the *input* dependence which occurs across writes. This is not really a dependence since reordering the reads will not cause problems[1]. Input dependences are sometimes used with attempting to analyze data reuse [85, 139]. In practice, compilers will conservatively assume that a dependence exists regardless of whether a value different than what is already in the array

---

[1]This statement may not be true for multi-threaded programs, but the current discussion concerns single-threaded programs. The multi-threaded case will be discussed in Section 2.6

```
           for (i = 1;  i < n;  i++) {
S1             b[2 * i] = …;
S2             … = b[2 * i − 1];
           }
```

(a) A loop with no data dependences.

```
for (i₁ = 1;  i₁ < n;  i₁ ++) {
    for (i₂ = 1;  i₂ < n;  i₂ ++) {
S1      c[2 * i₁, i₂] = d[i₁, i₂];
S2      d[i₁, i₂] = c[2 * i₁ − 2, i₂ + 1];
    }
}
```

(b) A loop with a dependence on the references to the c and d array.

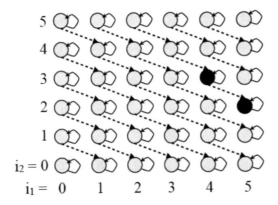

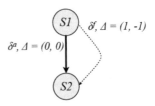

(c) An iteration space diagram (ISD) for the loop of (a). Dashed edges correspond to the dependence on c, and thicker edges correspond to the dependence on d.

(d) The dependence graph for the loop of (a). Dashed and thicker edges correspond to the same edges in the ISD of (a).

Figure 2.5: An example of a loop with multiple dependences.

element is written. We note that of these three kinds of dependences, anti and output dependences result from reusing storage, and only the flow or true dependence is essential to the computation, hence its name. In Chapter 4, we discuss transformations that can sometimes eliminate anti and output dependences, and that enable dependences to be ignored. Because the relative order of reads is immaterial when parallelizing an initially sequential program we will discuss input dependences no further.

Dependence analyses implicitly treat both a D_λ deep loop nest and a D_A dimensional array as a D_λ dimensional Cartesian space in the integers. Figure 2.5(b) shows a pair of references in a doubly nested loop. The iteration space diagram (ISD) in Figure 2.5(c) shows the Cartesian space that describes the iteration space of the loop and illustrates the dependence structure of the program at run time. Imposed on this Cartesian space are edges that relate points in the iteration space that access the same element of an array. In the iteration space diagram (ISD) of Figure 2.5(c) we see the edges corresponding to the accesses to the references to the c array in $S1$ and $S2$. For example, when $i_1 = 4$, $i_2 = 3$, element c[8, 3] is written, and when $i_1 = 5$, $i_2 = 2$ element c[8, 3] is read. That the same element is accessed, and the access is a write, creates a *dependence* on the c array between iterations $i_1 = 4$, $i_2 = 3$ and $i_1 = 5$, $i_2 = 2$. The dependence edge from $i_1 = 4$, $i_2 = 3$ to $i_1 = 5$, $i_2 = 2$ in the ISD represents this graphically, with the darkened nodes being those involved in the dependence just described. Additional edges on the graph are shown corresponding to the other points in the iteration space on which dependences exist. Note that for accesses on the d array the accesses that read and write the same element of the array are always in the same element, and so edges representing these are from an iteration to that iteration. We note that in the literature ISDs are sometimes shown with a node for each statement or reference in each iteration.

In Figure 2.5(d), we see the dependence graph for the loop nest of Figure 2.5(b). Nodes in the graph represent statements in the loop nest, and edges represent dependences. Edges are often labeled with the kind of dependence, in this case with δ^f to represent the flow dependence on c from $S1$ to $S2$ with a distance of $(1, -1)$, and δ^a to represent the anti-dependence on d with a distance of $(0, 0)$.

As seen in the ISD of our example, dependences always go forward in the execution of a program. A flow dependence represents a write to an array element followed by a read. An anti dependence represents a read to an array element followed by a write, and similarly for output and input dependences. Thus, the dependence is always from an earlier event to a later event. If the dependence is across iterations, then this means that the dependence moves forward in the iteration space as shown in the ISD. If the dependence is within an iteration, as is the dependence on d in Figure 2.5(b), then the dependence must be from an access that occurs earlier in the iteration to one that occurs later. Thus, for d, the read of d[i_1, i_2] occurs before the write of d[i_1, i_2], and therefore the dependence is an anti-dependence. This ordering is crucial because in part it is what allows the compiler to distinguish between a true dependence, and an anti dependence that it might be able to eliminate via code transformations.

When drawing dependences on an ISD or other diagram, the tail is always at the first action in the dependence, and the head at the second. The tail action is referred to as the *source* of the dependence, and the head action as the *sink* of the dependence. Thus, for a flow dependence the source is always a write and the sink is always a read, and for an anti-dependence the source is always a read and the sink is always a write.

The *direction* of a dependence tells how the iterations at the source and sink are related to one another. The direction is represented as a vector Γ, with $\Gamma[d]$, $1 \leq d \leq D$ containing the relationship between iterations at the source and sink (tail and head) of the dependence edge for the loop at nest level d in the loop nest containing the dependence. Let i_d be the value of the loop index variable i for the d'th loop in the loop nest for the dependence source, and i'_d be the value of the loop index variable for the dependence sink. The *distance* of the dependence is $(i'_d - i_d)$, and the direction is the sign of the distance, i.e., $\mathbf{sign}(i'_d - i_d)$. Thus, if the source is at an earlier iteration that the sink, the dependence direction vector element for loop i_d will be 1. It is possible for the direction to be positive, negative, or some combination of these, thus the direction for a loop i_d is represented as a set whose elements are any member of the power set of $\{1, 0, -1\}$, and the direction for the entire loop nest is a vector of these sets.

Since dependences must move forward in the iteration space of the loop, the question arises as to how the dependence distance and direction for some loops can represent a negative distance. Looking again at the loop of Figure 2.5, we see that the distance of the flow dependence on the read and write of c is positive on the i_1 loop, and negative on the i_2 loop. Thus, the dependence vector is $\Gamma = [\{1\}, \{-1\}]$. The reader can enumerate the iterations accessing the various elements of c on the read and write, and compare the i_1 and i_2 values when the same elements are read and written. The first non-zero element of a distance vector must be positive, and therefore the first non-zero element of the direction vector be lexicographically greater than $\{0\}$.

We note that sometimes a different notation for direction is used. In this notation, "$<$" is substituted for "1", "$>$" is substituted for "-1", and "$=$" is substituted for "0". The direction $\{<, =, >\}$ is sometimes represented as "*" in the literature. The advantage of this notation is that it lexically shows the relationship of the source and sink iterations (i.e., $i_d < i_d'$ for a positive distance), while the advantage of the 0, 1, -1 notation is that it represents the direction in a mathematically amenable notation.

A dependence for which the distance is non-zero, i.e., the direction contains a -1 or 1, is called a *loop carried* dependence because data is carried across iterations of the loop.

Logically, dependence testing has two phases. In the first phase, the variable references to be tested are determined, and then, in the second phase, the test for dependence is done. The dependence test driver determines pairs of references to be tested, and the direction vector under which they will be tested. It then calls a dependence test to actually perform the test.

2.3.1   DETERMINING REFERENCES TO TEST FOR DEPENDENCE

The dependence test driver operates over all pairs of accesses within some selected loop i, where i is typically the outermost loop in a loop nest that is being analyzed. The driver then examines all references which may access the same storage. Two references are considered to possibly access the same storage if (i) the references are to the same declared array, or (ii) the references are may or must aliased to the same array.

First, the relative *base addresses*, i.e., the relationship between the lowest elements of the array that can be accessed by each referenced array, is determined. In the easier (and common) case of references that are both to the same declared array, the base addresses must be the same. In the case of references that are possibly aliased to the same array, the situation may be more difficult, as shown in Figure 2.6. In the example, statements $S3$, $S4$ and $S5$ all pass the array z as arguments for the formal parameters a and b in the function leftShiftBtoA. In the case of the first call at $S3$, the statement at $S1$ essentially performs the operation

$$a[i] = a[i + 2]$$

for the iteration space from 0–999, and there exists an anti-dependence of distance "2" from b[i + 2] to a[i]. In the case of the second call at $S4$, the statement at $S1$ essentially performs the operation

$$a[i] = a[i - 1]$$

for the iteration space from 0–999, and there exists a flow dependence of distance 1 from a[i] b[i+1]. In the case of the third call at $S5$, the statement at $S1$ essentially performs the operation

$$a[i] = a[i + 500]$$

for the iteration space from 0–499, the regions accessed by the left-hand reference and the right-hand reference do not overlap, and there is no dependence.

```
        void leftShiftBtoA(double a[n], double b[n], int n) {
            for (i = 0; i < n; i ++) {
S1              a[i] = b[i + 1];
            }
        }
        ...
        void foo(...) {
S2          double z[1000];
S3          leftShiftBtoA(z[0], z[1], 1000);
S4          leftShiftBtoA(z[0], z[-2], 1000);
S5          leftShiftBtoA(z[0], z[499], 500);
            ...
        }
```

Figure 2.6: An example of issues raised in dependence analysis by aliased arrays. Different parameter values for a, b and n in function leftShiftBtoA lead to different types of dependence.

A compiler, when confronted by aliased references with non-equal base addresses, can either conservatively assume that flow and anti dependences exist between every pair of read and write references, and that an output dependence exists between every pair of write references, or the compiler can attempt to incorporate the offsets of the bases into the dependence equation. In the former case the dependence is assumed to exist over all possible valid directions, and in the latter case the test can proceed as normal with an extra offset term to augment the offset information provided by the subscript expression.

In the simpler case of common base addresses, or after an offset has been determined for the more complicated case of different base addresses for the array, the dependence tester needs to be called with different direction vectors to determine under what conditions the dependence exists. Given that there are $O(3^n)$ combinations of direction vectors to test under for an n deep loop nest, it is desirable that these tests be performed in a systematic way. In [39] a hierarchy of calls to the dependence testing routine under different direction vectors that allowed the search space to be systematically trimmed was proposed. Figure 2.7 illustrates this for a two deep loop nest. The test is first called on the subscripts with a direction of $(*, *)$ (i.e., any direction). If a dependence is possible under these directions, the test is called for the three directions at the next level $((<, *), (=, *)$ and $(>, *))$. Again, for any of the directions for which a dependence might still exist, the test is called for the next level of directions, as shown in the Figure 2.7. The observant reader will notice that the test on directions with a leading ">" direction, or with an "$(=, >)$" appear illegal because the dependence must move forward in the iteration space. If the test is performed with a source of S_i and a sink of S_j, the existence of a dependence with these directions implies a dependence from S_j to S_i with ">" directions being changed to "<", and "<" directions being changed to ">", yielding a dependence and direction that moves forward in the iteration space. This allows for a simultaneous testing of anti and flow dependences, and of output dependences with either statement being the source or sink.

We note at this time that the selection of references to test is typically done in a flow insensitive manner. That is, information about flow of control between the references is typically not used to determine if a dependence exists. Instead, all possibly aliased array references within the loop nest being analyzed are tested. There has been work in the field of software engineering to use control information in conjunction with dependence information to better understand the part of a program that is responsible for a the current value of a variable (i.e., the *slice* [241] for the variable) [212], but this topic is beyond the scope of this lecture.

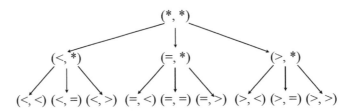

Figure 2.7: The hierarchy and order of testing on different direction vectors.

2.3.2   TESTING FOR DEPENDENCE

The test for dependence takes as input four items, listed below along with an example of each item from Figure 2.8.

1. The subscript expressions for the two references being tested, e.g.,

$$\begin{aligned} \sigma_1(I) &= a_1 * i_1 + a_2 * i_2 + a_3 * i_3 + a_4 * i_4 + a_0 \text{ and} \\ \sigma_2(I') &= b_1 * i'_1 + b_2 * i'_2 + b_5 * i'_5 + b_0 , \end{aligned}$$

where the as and bs represent loop invariant values. The subscripts are assumed to be linear functions in several dimensions, i.e., *affine* functions;

2. the bounds for each loop index i_d in the common loop nest surrounding the references (if not available the maximum allowed value (e.g., `MAXINT`) for the bounds can be used), e.g.,

$$L_I = [L_{i_1}, L_{i_2}, L_{i_3}, L_{i_4}, L_{i_5}],$$
$$U_I = [U_{i_1}, U_{i_2}, U_{i_3}, U_{i_4}, U_{i_5}] ;$$

3. a direction for each variables in the loop nest(s) surrounding the loop, (e.g., $\Gamma = [\gamma_1, \gamma_2, \gamma_3, *, *, *]$); and

4. the non-common loops (denoted by i_*) surrounding each reference, e.g., $i_* = [i_3, i_4], i'_* = [i_5]$.

The dependence tester now needs to answer the question:

Given subscript functions $\sigma_1(I), \sigma_2(I')$ on the index variables I and I' of the loops surrounding each reference, does an integer solution exist for the equation $\sigma_1(I) = \sigma_2(I')$ (the references access the same elements of the array) subject to the constraints that:

1. $\forall i_d, 1 \leq d \leq D$ in the D-deep common loop nest $i_d \gamma i' == $ true, $1 \leq d \leq D$ (i.e., the direction vector is satisfied by the solution);

2. $L i_d \leq i_d \leq U_{i_d}, 1 \leq d \leq D$ (i.e., the solution is within the loop iteration space); and

3. $lb_d \leq i_d \leq ub_d, i_d \in i_*$ and $L_{i_d} \leq i'_d \leq U_{i_d}, i'_d \in i'_*$ (i.e., the solution for the non-common loop indices is within the loop iteration space of the non-common loops).

```
         for (i1 = lb1; i1 < ub1; i1 ++) {
           for (i2 = lb2; i2 < ub2; i2 ++) {
             for (i3 = lb3; i3 < ub3; i3 ++) {
               for (i4 = lb4; i4 < ub4; i4 ++) {
S1                z[a1 * i1 + a2 * i2 + a3 * i3 + a4 * i4 + a0] = ...;
               }
             }
             for (i5 = 1; i5 < ub5; i5 ++) {
S2              z[b1 * i1 + b2 * i2 + b5 * i5 + b0)] = ...;
             }
           }
         }
```

Figure 2.8: An example of references with non-common loops (those with italicized *for*s) in their surrounding loop nests.

The first part of this question is answered using the *gcd test*. In the gcd test, the Diophantine equation for the subscripts is formed:

$$
\begin{aligned}
b_n - a_m \;=\; & a_1 * i_1 - b_1 * i_1' + a_2 * i_2 - b_2 * i_2' + \ldots + \\
& a_D * i_D - b_D * i_D' + a_{D+1} * i_{D+1} + \ldots + \\
& a_n * i_n + b_{n+1} * i_{n+1}' + b_{n+2} * i_{n+2}' + \ldots + b_m * i_m'.
\end{aligned}
$$

A well-known result in number theory says that a Diophantine equation has a solution if and only if the gcd of the coefficients evenly divides the constant right-hand side, i.e., that there is a solution only if

$$
0 = b_0 - a_0 \bmod \gcd(a_1, a_2, \ldots, a_n, b_1, b_2, \ldots, b_D, b_{n+1}, b_{n+2}, \ldots, b_m) .
$$

That this is true can be shown using the simpler equation

$$
a_0 = a_1 \cdot i_1 + a_2 \cdot i_2 + \ldots + a_n \cdot i_n
$$

and observing that if

$$
g = \gcd(a_1, a_2, \ldots, a_n)
$$

then

$$
g \cdot a_0' = g \cdot (a_1' \cdot i_1 + a_2' \cdot i_2 + \ldots + a_n' \cdot i_n)
$$

and therefore, if the gcd of the coefficients does not evenly divide the constant term, there can be no solution, because if the right and left-hand side are equal they must have the same divisors.

If there are multiple subscript expressions, as in the case of multi-dimensional arrays, then there are three possible approaches. First, the equations for each dimension can be formulated and solved independently. If any equation does not have a solution, dependence is disproved. Second, the subscripts expressions can be combined by linearizing them. These subscript expressions are combined in essentially the same way as they are combined by a compiler generating code for a multi-dimensional array access. These new subscript expressions are then equivalenced and the resulting equation is checked for a solution. Third, the *extended gcd test* can be used, with the equations for all of the dimensions being solved simultaneously. This is discussed in more detail in Chapter 7.

Unfortunately, coefficients of index variables are often 1, and thus the gcd is often 1. Thus, it is essential to determine if the solution to the Diophantine equation is within the loop bounds and consistent with the direction vector. A precise answer to whether the solution fits these constraints requires linear programming, which has exponential complexity. This has motivated the development of various heuristics. Banerjee's inequalities were the earliest such heuristic. They attempt to find an upper and lower bound on the right-hand side of the equation. Consider a term $a * i - b * i'$ in the right-hand side. If the direction vector is 1, then $i < i'$, and thus the lower bound of the term (assuming a, b positive) is $a * lb_i - b * ub_i$. Similar expressions can be derived for the case of $a > 0, b < 0; a < 0, b > 0$, and $a < 0, b < 0$. From these the minimum (L) and maximum (U) values of the dependence equation can be found, and by the mean value theorem, if $L \leq b_0 - a_0 \leq U$, the equation is assumed to have a solution in the iteration space, and a dependence is assumed to exist. If it is not the case that if $L \leq b_0 - a_0 \leq U$ then dependence has been disproved. We note that in rectangular loop nests (i.e., nests in which the upper and lower bounds of all loops are constants within the iteration space) with known loop bounds, this test is precise [22].

A tighter bound can be found by using the *Power Test* [252]. When solving a set of Diophantine equations using the extended gcd test the solutions for each index variable can be expressed in terms of parametric equations and parametric variables t. Thus, it may be that $i_1 = t_2 + 1$ and $i_1' = t_2$. If $i_1 < i_1'$, because of the direction vector under which the test is applied, then it must be that $t_2 + 1 < t_2$. If $i_1' > 0$, then $t_2 > 0$, and $t_2 + 1 > t2$, thus the solution to the system of equations is inconsistent with the direction vector, and the solution is inconsistent. Fourier-Motzkin elimination gives a systematic way of finding these inconsistencies. The Power Test can be exponential in the number of index variables (i.e., depth of the loop), but provides faster solutions than other techniques [119, 134] that utilize Fourier-Motzkin elimination to prove independence. When originally developed it was too slow for general use, but with modern machines can be applied to those references that may be dependent after applying Banerjee's inequalities. The Power Test is exact for loops whose bounds are affine functions of outer loop indices, and in which all constants in the dependence equations and loop bounds are known.

The Omega test [184] is a powerful technique that uses an extension of Fourier-Motzkin Elimination to determine if a solution exists to a dependence equation. The Omega test is exponential in the number of index variables. The Omega tests can solve for solutions in some non-linear dependence equations. Of particular importance are those that arise from index expressions of the form $c \cdot i + n$, where the value of n is unknown at compile time. This is done by adding the variable in the index expressions with an unknown value into the system to be solved as a new variable. As well, equations involving integer division and integer remainder operations can be handled. This is done by adding n as another variable to the system being solved. In [181, 182] it is claimed that in all situations where other dependence tests based on Fourier-Motzkin elimination and simpler analyses are accurate, the Omega test is sufficiently fast, especially when a careful implementation is done.

Using *hybrid analysis*, described in [202], conditions that lead to dependence not being disproved are identified, and tested at run time. Consider the pair of references $a[i]$ and $a[i + N]$. If $N = 0$ then no loop carried dependence exists, i.e., no dependence that crosses iterations of a loop. If $N > 0$ the dependence source is $a[i]$ and its sink is $a[i + N]$. Depending on the lexical relation of the references in the program, it might be possible to place a test at run time to evaluate the value of N, and if $N = 0$, the loop could be run in parallel, and if $N > 0$ or $N < 0$ then it might be possible to apply loop fission to the loop (see Section 5.3) and then parallelize the loop.

2.3.3   ARRAYS OF ARRAYS AND DEPENDENCE ANALYSIS

Some languages, most prominently Java, represent all multi-dimensional arrays as "arrays of arrays". Thus, in Java, an $m \times n$ two-dimensional array consists of a one-dimensional array with m-elements that are pointers to arrays (the *pointer array*), and $r \leq n$ one-dimensional arrays representing the rows of the array, as shown in Figure 2.9. There are numerous problems associated with analyzing such arrays [155, 159], but we will only consider those raised by dependence analysis.

As shown in Figure 2.9, it is possible for a row array to be addressed by multiple elements in the pointer array. As a consequence, of this, showing that two references never access an element with the same coordinates is insufficient to show that they are independent. Thus, while the subscript expressions of the reads and writes of a never have the same values, there is still a flow dependence with distance zero on the i loop.

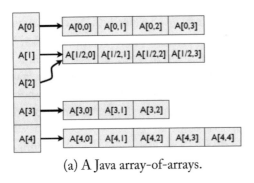

(a) A Java array-of-arrays.

$$
\text{for } (i = 0; i < n; i++) \\
a[2*i] = a[2*i-1];
$$

(b) The code used to create the array of (a).

Figure 2.9: An example of a Java array of arrays.

2.4   CONTROL DEPENDENCE

When a compiler transforms a program such that the order, kind of operation, or the conditions under which an operation is executed change, it must ensure that the output of the program is as if

1. the inputs to the operation are unchanged;

2. that the outputs of the moved operation do not change the input to other dependent operations;

3. that the operation is always performed at run time when it might have been performed during the original program; and

4. that the operation is never performed at run time when it would not have been performed during the original program.

The first two conditions are determined by the data dependence structure of the program. The final two constraints are determined by the *control dependence* [62, 75] structure of the program. If a control dependence exists from a source statement S_{so} to a sink statement S_{si}, then whether or not some instances of S_{si} execute is dependent on the outcome of S_{so}.

Consider the program and CFG of Figures 2.10(a) and (b). In Section 1.3.3 we discussed the concepts of dominance and dominance frontier, and how they relate to finding join points in a program. Determining control dependences solves a related problem. The statements in some block b_{si} are control dependent on the last statement in a block b_{so} if the block b_{so} causes a fork rather than a join in the CFG, and which path is taken at the fork determines if some instance of b_{si} is executed. It is precisely these forks that constitute the sources of control dependences.

There are two well-known ways of finding control dependences. The first is from [75], and there the following condition is given for control-dependence to exist between two nodes N_{so} and N_{si} in the forward CFG. N_{si} is control dependent on N_{so} if:

1. there is a path from N_{so} to N_{si} such that additional node(s) N on the path must be post-dominated by N_{si};

2. N_{so} is not post-dominated by N_{si}.

Intuitively, N_{si} is control dependence on N_{so} if all nodes on the path between N_{so} and N_{si} can reach N_{si} (the first condition), but N_{si} doesn't have to (the second condition), and all other nodes N between N_{so} and N_{si} on the path from N_{so} to N_{si} must pass through N_{si} to reach the exit of the CFG (the second condition).

Another technique, related to dominance frontiers, is given in [62]. If the *reverse CFG* is formed (shown in Figure 2.10(c)), i.e., a CFG where every edge $b_i \rightarrow b_j$ is replaced with an edge $b_j \rightarrow b_i$, then forks in the graph become joins, and the joins become forks. Thus, by finding the dominance frontier of a block b_{si} in the reverse CFG, the fork points nearest to b_{si} in the regular CFG are located, and hence the sources b_{so} of all control dependences $\delta^c(b_{so}, b_{si})$ involving b_{si}. Figure 2.10(d) shows the control dependence in the program.

The *program dependence graph* [75] displays both the data and control dependences found in a program. By showing only the dependence edges, all edges not ordered by a dependence edge can be executed in parallel. The PDG for the program of Figure 2.10(a) can be found in Figure 2.11.

Control dependences can be essentially converted to data dependences [7], allowing a common framework to be used. Figure 2.12 shows a `for` loop with an `if` statement. The technique first converts the `if` to an assignment into a vector of bits, and then tests the appropriate bit before executing each statement that is under the `if`. Dependence testing will reveal the appropriate orderings that must be enforced, and can provide information about the iteration for the control dependence source.

2.5   USE-DEF CHAINS AND DEPENDENCE

The dependence test described above is typically applied to references that have at least one loop in common in their enclosing loop nests. It is often useful to determine the sources of read values on

```
    ...
S1   a[n] = b[n] + c[n]
S2   for(i = 0; i < n; i++){
S3      b[i] = ...
S4      if(i − 2 < 0)
S5          b[i − 1] = a[i] + c;
        else
S6          a[i] = a[i] + b[i − 2];
S7   }
```

(a) A program.

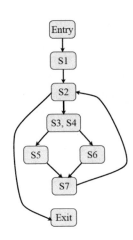

(b) The control flow graph (CFG) for the program of (a).

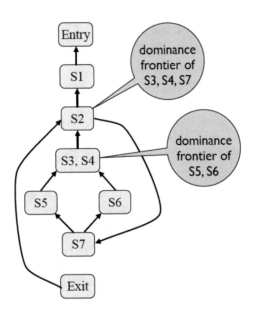

(c) The reverse control flow graph for the program of (a).

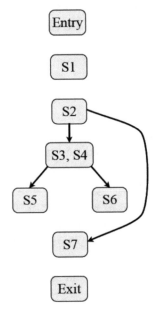

(d) The control dependences in the program.

Figure 2.10: Example of a control dependencies.

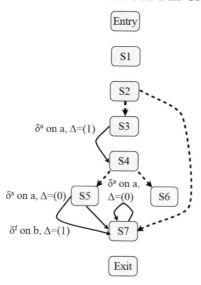

Figure 2.11: The program dependence graph (PDG) for the program of Figure 2.10(a), where solid edges represent data dependences and dashed edges represent control dependences.

```
for (i = 0; i < n; i ++) {          for (i = 0; i < n; i ++) {
    if (a[i] > 0)                       cond[i] = a[i] > 0
        a[i] = a[i + 1] * a[i];         if (cond[i]) a[i] = a[i + 1] * a[i];
        b[i] = b[i + 1];                if (cond[i]) b[i] = b[i + 1];
}                                   }
```

Figure 2.12: An example of converting control to data dependences.

scalars and on array elements when there is no common loop nest, such as is shown in Figure 2.13. Whether the references access the same array elements can be determined by applying the gcd test to the subscripts. Since the reference in $S1$ will execute first, we know that the dependence source is $S1$ and the sink is $S2$. Since there are no common enclosing loops, the direction and distance vectors have no elements.

$$
\begin{array}{ll}
      & \text{for (i = 0; i < n; i ++) \{} \\
S1   & \quad \text{a[i] = ...;} \\
      & \} \\
      & \text{for (i = 0; i < n; i ++) \{} \\[6pt]
S2   & \quad \text{... = a[i + 2] = ...;} \\
      & \}
\end{array}
$$

Figure 2.13: Array accesses without a common loop nest.

2.6   DEPENDENCE ANALYSIS IN PARALLEL PROGRAMS

An analysis analogous to dependence analysis can be applied to identify orderings in multithreaded programs that must be honored by both hardware and compiler transformations. This analysis is necessary when the program being analyzed is already parallel. Before considering the case of dependence analysis in parallel programs we will revisit what it means for there to be a dependence in a sequential program.

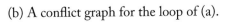

```
        for (i = 0; i < n; i++) {
S1        a[i] = ...
S2        ... = a[i − 1]
        }
```

(a) A loop with a flow dependence from $S1$ to $S2$.

(b) A conflict graph for the loop of (a).

(c) The conflict graph of (b) showing an orientation of the conflict that gives a legal execution of the program.

(d) The conflict graph of (b) showing an orientation of the conflict that gives an illegal execution of the program.

Figure 2.14: An example of viewing a dependence as an edge in a conflict graph.

In Figure 2.14(a) a sequential program with a loop containing a loop carried dependence (a flow dependence from $S1$ to $S1$ with a distance of 1) is shown. Let us consider an alternative way of viewing this dependence by building a *conflict graph*. In the conflict graph, nodes will be statements or references, and edges will represent either (i) orders implied by the program semantics (*program edges*) or (ii) a pair of accesses to the same memory location, where at least one is a write (*conflict edges*). A conflict graph for the loop body of Figure 2.14(a) is shown in Figure 2.14(b), and the program edge is shown as a solid line and the conflict edge as a dashed line.

Now consider the execution order of the references at each end of the conflict. If the reference that executes first is the reference that would have executed first if all program edges were honored (as seen in the orientation of the conflict edge in Figure 2.14(c)) then the value read by the reference to a in $S2$ will be the value written earlier by a write to a in $S1$, and the execution of the loop would be correct. If, however, the read of $S2$ executed before the write in $S1$ for one or more conflicting accesses of a, the orientation of those conflict edges would be as shown in Figure 2.14(d), and the outcome of the loop execution would be incorrect. That this is the case can be seen by observing that there is a cycle in the conflict graph of Figure 2.14(d), the cycle involves both program and conflict edges, and because there is this cycle the orientation of the conflict edges is inconsistent with the order of accesses implied by the program edges. It is the inconsistency that leads to an error in the execution.

Dependence analysis is looking for similar inconsistencies in a different way. The gcd test shows that there is a conflict between two references, since two references can access the same memory location. Testing under the sign determines if the possible source reference can access the memory location before the possible sink reference, a dependence exists, and the cycle corresponding to this dependence can be found by orienting the conflict edge to run from the sink to the source of the dependence.

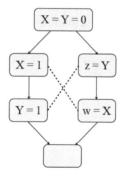

```
X = Y = 0;
cobegin
    X = 1;
    Y = 1;
//
    z = Y;
    w = X;
end cobegin
```

(a) An explicitly parallel program.

(b) Dependences (dark solid lines) in an explicitly parallel program. Enforcing the dependences allows only sequentially consistent executions, and disallows the outcome "$v = 1, w = 0$"

Figure 2.15: An example of an explicitly parallel scalar program.

Testing for dependence in explicitly parallel programs requires looking for similar inconsistencies between possible conflict edge orientations and orders implied by the parallel program semantics. We now discuss how this is done, using the results of Shasha and Snir [209].

Figure 2.15(a) shows a fragment of an explicitly parallel program, and Figure 2.15(b) shows the conflict graph for that program. The cobegin construct defines different threads, with each thread being the code that is delimited by the cobegin statement and the "//" markers. The conflict graph for the parallel program has solid lines that indicate *program orders*, i.e., orders that are defined by the program execution order (in this case, for a sequentially consistent program execution order)

```
a[1 : n] = 0;
parfor (i = 1;  i < n;  i++) {
    a[i] = b[i − 2];
    b[i] = a[i − 3];
}
```

(a) An explicitly parallel program with arrays that are shared across threads.

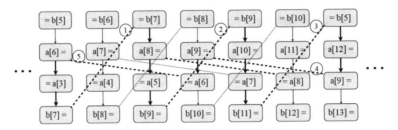

(b) The conflict graph for the program of (a) showing program edges (solid light lines), conflicts (dashed lines) and dependences resulting from inter-thread analysis (bold solid lines). Numbers in the circles indicate the order that edges can be followed to form a cycle.

Figure 2.16: An example of an explicitly parallel program with arrays.

and dashed lines that indicate *conflicts* among accesses in the different threads. As mentioned above, a conflict can be conservatively defined as two accesses to the same storage location, at least one of which is a write. The more precise definition adds the proviso that the order that the accesses occur in can be determined by examining the state of the program, which in practice means that writes place a different value into the storage location than what was already there.

Some execution orderings of accesses at the endpoints of conflict edges can lead to program outcomes that are inconsistent with those that are possible if the program orders are followed. These orders are exactly those orders that lead to cycles in the conflict graph. Thus, in Figure 2.15(a), if z = Y executes after Y = 1 its value will be 1. Because of the program edge from X = 1 to Y = 1, if z = Y executes after Y = 1 it should also execute after X = 1. And because of the program edge from z = Y to w = X, w = X should execute after z = Y, and transitively should execute after X = 1. Therefore, if z has the value of 1 then w should also have the value of 1. If w has a value of 0, it implies that somehow w = X executed before X = 1 in the same execution that z = Y executed after Y = 1. Tracing the corresponding conflict orientations on the conflict graph yields a cycle.

An important result of [209] is that enforcing all program edges involved in these cycles will prevent the invalid orientations from occurring at run time, and will prevent invalid executions.

Thus, the program edges involved in cycles can be thought of as dependences in the explicitly parallel program.

```
X = Y = 0;
cobegin
    X = 1;
    Y = 1;
//
    z = Y;
    w = X;
//
    ...= z;
    ...= w;
end cobegin
```

(a) A program whose conflict graph has minimal and non-minimal mixed cycles.

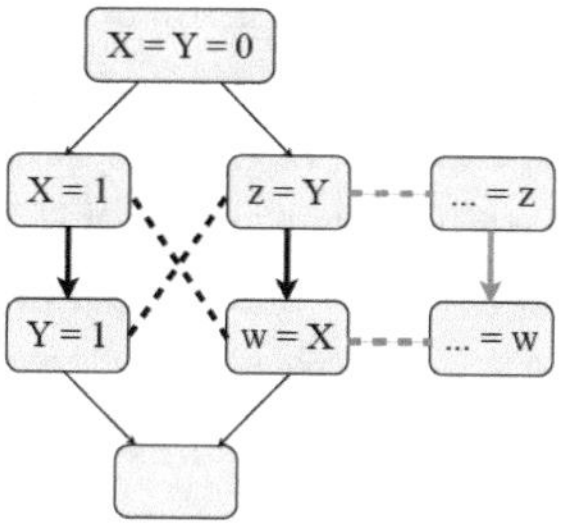

(b) A conflict graph with a minimal and non-minimal mixed cycles.

Figure 2.17: An example of a conflict graph with a minimal cycle involving the nodes X = 1, Y = 1, z = Y, and w = X; and a non-minimal cycle involving the same nodes as well as ...= z and ...= w.

In [209], Shasha and Snir prove that dependences in parallel programs can be identified by identifying all of the *minimal, mixed cycles* in a conflict graph for the program, and marking all program edges in these graphs as dependences. A mixed cycle is a cycle that contains both conflict and program edges. This is necessary because a cycle without conflict edges does not capture the state resulting from reads and writes occurring out of order, and a cycle without program edges does not expose an inconsistency with respect to the ordering semantics of the program. A minimal cycle is one that contains no smaller mixed cycle that spans two or more threads. An example of a minimal mixed cycle, and a non-minimal mixed cycle are shown in Figure 2.17(b). This edges of the minimal mixed cycle are shown bold and dark, and the additional edges in the non-minimal cycle (involving nodes "... = z" and "...= w" are shown in grey.

When a compiler analyzes the threads (and functions in a thread) independently of other functions and threads, alias analysis will show that X and Y in the program access separate storage and are independent, and therefore in an analysis that assumes that the program is sequential they may be re-ordered by either the hardware or the compiler. Either reordering the writes or the reads of X and Y will lead to the disallowed execution.

Figure 2.16(b) shows the conflict graph for the program of Figure 2.16(a). This program fragment contains a parallel parfor loop that has cross iteration conflicts. Again, the bold edges are involved in a cycle, and again this cycle indicates that an execution that is not consistent with the program orders can result if program edges involved in the cycle are not enforced. Conflicts can be detected by applying a dependence test with a direction vector of "*" elements to potentially conflicting references.

In [209] it is argued, but not proven, that a polynomial number of such cycles exist. It is shown, however, in [128] that finding the cycles is exponential in the number of threads, which makes this algorithm extremely expensive for compilers. Yelick and Krishnamurthy show in [128] that for SPMD programs the number of threads can be modeled with two threads, without loss of precision (compared to doing the analysis on a conflict graph from [209]), and leads to an $O(|T|^2)$ solution, where $|T|$ is the number of threads. In Figure 2.19 an example of how SPMD programs can be handled is shown. In the graph the program of 2.16 each iteration is potentially a thread, and therefore up to n threads might need to be represented on a conflict graph as defined in [209]. Yelick and Krishnamurthy "roll" all of these threads into a *left* and a *right* pair of threads. Each conflicting pair of nodes (v_p, v_q) leads to conflicts between the corresponding v_p in the left (right) thread, and v_q in the right (left) thread. All program edges in the original conflict graph are placed on the right thread of this graph.

To find a program edge that is in a critical cycle, i.e., a program edge that must be enforced, the algorithm selects a pair of nodes (v_p, v_q) from the left side. It then attempts to find a path involving both conflict and program edges from v_p to v_q. If such a path exists, the edge (v_q, v_p) must be honored. Thus, the ordering from "a=" to "=b" must be honored because there is a path from "b=" in the left thread through the nodes "b=", "=a", "a=" in the right thread, and then to "a=" in the left thread.

In Figure 2.19(c), the same program as shown in Figure 2.19(c) is given, except now the references to the a and b arrays are subscripted. As can be seen from the graph of Figure 2.19(d) no cycles exist, but because the methods of both Shasha and Snir and Yelick ignore subscripts, both will still find a cycle. The algorithm of Midkiff [152] partially overcomes this limitation by finding cycles in parallel loops with constant distance dependences. The paper makes the observation that an edge on a conflict with a distance k moves from an iteration i to an iteration $i \pm k$, depending on which direction in the iteration space (towards higher or lower iterations) the edge is traversed. By using Kirchoff's laws of flows, the relative number of edges that can be traversed through the graph can be computed, and from this and the edge distances a linear programming problem can be solved that says whether it is possible to travel from an iteration back to that iteration on the conflict edges, and therefore whether or not a cycle exists.

Figure 2.18 shows an example of this (note that for clarity transitive program edges are not shown). Each mixed cycle in the graph is enumerated, and considered in turn. For example, with the cycle $b[1_3] \rightarrow b[j_4] \rightarrow b[j_1] \rightarrow b[i_2] \rightarrow b[j_3]$, the set of constraints shown in Figure 2.18(c) is formed. Each program edge contains a relationship between the index variable values at the head and tail of the program edge, and these are added to the system. Each conflict edge is also labeled with a relationship between the index variable values at the tail and head of the edge, thus for the edge $(b[j_4], b[i_4])$ the relationship is $i_4 = j_4$. For clarity, this labeling has only been shown on the very bottom conflict edges $((b[j_4], b[i_4])$ and $(b[i_4], b[j_4]))$. Finally, each edge is labeled with a value e_i. This represents the number of times this edge can be traversed, and the constraints on each e_i are determined by Kirchoff's Laws of Flow. Thus, in our example, the sum of the number of times edges

```
parfor (i = 0; i < n; i ++) {
    ...= a[i]
    a[i] = ...
    ...= b[i]
    b[i] = ...
}
```

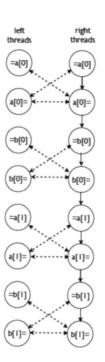

(a) The explicitly parallel program of Figure 2.19(b).

(b) A graph used to perform a subscript aware analysis to find critical cycles.

$$
\begin{aligned}
i_2 &\leq i_3 \\
j_4 &= i_3 \\
j_1 &= i_4 + e_8 \\
i_2 &= j_1 \\
e_1 1 &\leq e_8 \\
e_1 6 &\leq e_2
\end{aligned}
$$

(c) The system of equations that arises from determining if there is a cycle $b[1_3] \rightarrow b[j_4] \rightarrow b[j_1] \rightarrow b[i_2] \rightarrow b[j_3]$.

Figure 2.18: An example of using SPMD program graphs to analyze explicitly parallel programs with SPMD parallelism. Dashed lines indicate conflicts, solid lines program flow.

```
parfor (i = 0; i < n; i ++) {
    ...= a
    a = ...
    ...= b
    b = ...
}
```

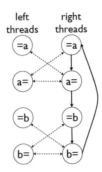

(a) An explicitly parallel program.

(b) The SPMD program graph for the program of (a).

```
parfor (i = 0; i < n; i ++) {
    ...= a[i]
    a[i] = ...
    ...= b[i]
    b[i] = ...
}
```

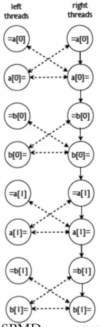

(c) An explicitly parallel program with subscripted references.

(d) The SPMD program graph for the program of (b).

Figure 2.19: An example of using SPMD program graphs to analyze explicitly parallel programs with SPMD parallelism. Dashed lines indicate conflicts, solid lines program flow.

e_{11} and e_5 are traversed cannot be more than the number of times node $a[j_1]$ is entered, thus we know that both $e_{11} + e_5 < e_4$ and $e_{11}, e_5 \leq e_4$. Moreover, it is known that every edge is traversed at least 0 times, i.e., $e_p \geq 0$, and that edges in the tested cycle must be traversed at least once, and so all $e > 1$ for these edges. Combining these constraints involving the edges in this example yields the set of equations of Figure 2.18(c). A linear programming solver can be used to see if a solution exists, and if it does the cycle is assumed to exist, and program edges in the cycle must be enforced.

In [225] a heuristic that finds cycles in non-SPMD programs is presented that gives good results. This technique works by checking for necessary, but not sufficient, conditions for a minimal mixed cycle to exist. Given a conflict graph, their technique picks a pair of nodes A and B joined by a program edge (A, B). We will demonstrate the heuristic using the program of Figure 2.15(a), and will test to see if there is a minimal mixed cycle involving the program edge going from the node X = 1 (A) to Y = 1 (B). Two other nodes are identified, which we will call C and D^2. Let D be the node z = Y and C be the node $w = X$. Then a minimal mixed cycle may exist if:

1. a conflict edge exists between A (X = 1) and C (w = X);

2. a conflict edge exists between B (Y = 1) and D (z = X);

3. A (X = 1) and C (w = X) may occur in different threads;

4. B (Y = 1) and D (z = Y) may occur in different threads;

5. A (X = 1) may occur after C (w = X) and D (z = Y);

6. B (X = 1) may occur before C (w = X) and D (z = Y);

7. D (z = Y) may occur before C (w = X).

The first four conditions determine that there are conflict edges from the endpoints of the program edge being tested to accesses in different threads. The last condition says that a path of program edges might go from D to C. This combined with the existence of the (A, B) program edge and the two conflict edges establishes that there may exist a path from B to D to C to A and then back to B, which establishes the cycle. Finally, the fifth and sixth conditions establish that the executions of A and B, and C and D can overlap and interfere with each other, and therefore may lead to a dependence. In [225] it is shown that this test eliminates the majority of the dynamic program edges that must be enforced in the benchmarks studied.

In most languages with support for parallel and concurrent programming, different threads (as opposed to different instances of a single static thread) are activated by executing different functions or methods. This means that the analysis to determine dependences in explicitly parallel programs is inherently inter-procedural. The language with the best defined consistency model[3] is

---

[2]In the paper they are called X and Y, but to avoid confusion with the variables in the example we have renamed them.
[3]Although with recent work on the C, C++ and OpenMP consistency models, it may not be unique at the time this lecture is being read

currently Java, and its dynamic compilation model precludes, for performance reasons, analyses that require inter-procedural analysis. As a consequence, of this, and the lack of (or only recent definition of) well-defined memory models in other languages, these techniques have to our knowledge only been implemented in research compilers, with one exception. That exception is Unified Parallel C (UPC) [237], which used the analysis of [128] to optionally support a sequentially consistent execution model.

CHAPTER 3

Program parallelization

This chapter discusses the use of dependence information to guide automatic program parallelization. We focus on loops without data dependences that would prevent any parallelization of the loop. In practice, it is often necessary to perform additional transformations to allow a loop to be partially or completely parallelized. Transformations to eliminate dependences, or to allow them to be ignored, are discussed in Chapter 4. Other transformations that may enhance parallelization are discussed in Chapters 5.

This chapter is organized as follows. In Section 3.1, the easy case of parallelizing loops with no dependences that prevent full parallelization of the loop is discussed. In Section 3.2, the parallelization of loops with more complicated dependence structures is discussed, and these capabilities are further expanded with a view towards targeting vector hardware in Section 3.3. Section 3.4 examines how producer/consumer synchronization (in contrast to *mutex* synchronization) can be used to extract more parallelism than would otherwise be possible from a loop. In Sections 3.5 and 3.6, issues arising from more complicated control structures, in particular from recursion and `while` loops, is discussed. Finally, Section 3.7 discusses *software pipelining*, which exposes fine-grained instruction level parallelism to the processor.

3.1 SIMPLE LOOP PARALLELIZATION

As discussed in Section 1.1, for a loop to be parallel without any additional transformations its iterations must be independent, i.e., there can be no flow of values or reuse of memory locations across iterations. More formally, there can be no loop-carried dependences on the loop to be parallelized, i.e., all dependences in the loop to be made fully parallel must have a direction of "0" ("="). As mentioned above, in some cases problematic dependences can be eliminated or ignored. It is also generally desirable that the parallel loop be the outermost loop in the nest, as this maximizes the amount of work done in the parallel loop, and minimizes the number of instances of the parallel loop that are started, and the ensuing overhead. Loop interchange (Section 5.5) accomplishes this.

When loop-carried dependences have non-equal directions on several loops, it is sometimes possible to parallelize an inner loop. More specifically, if all dependences that are loop-carried, i.e., that have a non-equal direction on the i_2 loop also have a non-equal direction on some outer loop(s) i_1, the inner i_2 loop can be parallelized. An example of this can be seen in Figure 3.1. In the example, the outer i_1 loop is forced to execute sequentially (i.e., is *frozen*) to enable the instance of the inner i_2 loop within an iteration of the i_1 loop to execute in parallel. By examining the order of execution of the different i_2 loop iterations it can be seen how freezing the i_1 loop enables the i_2

loop parallelization. Because the source of the dependence is in an earlier iteration of the i_2 loop than the sink is, and the i_1 loop is executing sequentially, the sink of the dependence must execute later than the source, even when the i_2 loop is parallel. This forces the dependence to be honored, allowing the i_2 loop to be executed in parallel.

```
for (i₁ = 0; i₁ < 4; i₁ ++){
    for (i₂ = 0; i₂ < 8; i₂ ++){
        a[i₁][i₂] = a[i₁][i₂] + b[i₁];
        c[i₁][i₂] = a[i₁ − 1][i₂ + 1];
    }
}
```

(a) A loop where freezing the outer i_1 loop exposes parallelism.

```
for (i₁ = 0; i₁ < 4; i₁ ++){
    parallel for (i₂ = 0; i₂ < 8; i₂ ++){
        a[i₁][i₂] = a[i₁][i₂] + b[i₂];
        c[i₁][i₂] = a[i₁ − 1][i₂ + 1];
    }
}
```

(b) The loop of (a) with the i_2 loop parallelized.

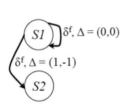

(c) The dependence graph for the loop of (a).

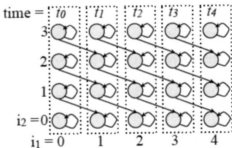

(d) An iteration space diagram showing the execution orders of different iterations and the dependence sources and sinks.

Figure 3.1: A example of how freezing outer loops can enable inner loop parallelization.

3.2 PARALLELIZING LOOPS WITH ACYCLIC AND CYCLIC DEPENDENCE GRAPHS

Substantial parallelism can often be extracted from loops whose dependence structure is more complicated than the loops discussed in Section 3.1. In particular, we discuss how to extract parallelism when the dependence graph may be cyclic and loop freezing cannot be used to break the cycles. The following steps will be performed.

1. Construct a dependence graph for the loop nest.

2. Find strongly connected components (SCCs) formed by cycles of dependences in the graph, contract the nodes in the SCC into a single large node.

3. Mark all nodes in the graph containing a single statement as parallel.

4. Topologically sort the nodes in the graph so that all inter-node dependences are lexically forward.

5. Group independent, unordered, nodes reading the same data and marked as parallel into new nodes to optimize data reuse.

6. Perform loop fission (see Section 5.3) to form a new loop for each node in the sorted dependence graph.

7. Mark as parallel all loops resulting from nodes whose statements are marked as parallel in the sorted graph.

These steps will be explained in detail by means of an example in the remainder of this section.

When the dependence graph for a loop is acyclic, as shown in Figure 3.2(a) (showing a program) and Figure 3.2(b) (showing the acyclic dependence graph for the program) the loop can be fully parallelized by breaking it into multiple loops, each of which has no loop-carried dependences. If none of these multiple loops executes until all dependences into it are satisfied (i.e., until after the loop that contains the source of the dependence has already executed), then the barrier that is, by default, at the end of every parallel loop (see Section 1.2.1) will force the writes from each instance of the inner parallel loop to complete before other instructions execute, and the dependence is honored. Thus, the goal of the compiler is to break up the loop with dependences into multiple loops, none of which contain the source and sink of a loop carried (cross-iteration) dependences and then to order these new loops so that they only execute when all of the dependences they are involved in have been satisfied.

To do this requires ensuring that all dependences are *lexically forward*, i.e., that in branchless code the sink of the dependence is lexically forward of the source of the dependence. This can be done by topologically sorting the dependence graph. Because the dependences form a partial order on the nodes of Figure 3.2(b) there are several possible orderings of the nodes resulting from a topological sort; one valid order is shown in the graph of Figure 3.2(c).

Once the dependence graph is topologically sorted the compiler can reorder the statements in an order that is consistent with the topological sort order, i.e., the loop is transformed by reordering the statements to match the topological sort order. *Loop fission* (Section 5.3), which is always legal with *lexically forward* dependences, is then performed, yielding the sequence of loops seen in Figure 3.2(d). In the transformed program, no loop has cross iteration dependences. A more efficient parallelization can be performed by a different partitioning of statements among loops that is still consistent with the ordering implied by the topological sort. The more efficient partitioning keeps statements that are not related by a loop-carried dependence together in the same loop. Doing this with the original graph yields an ordering $S2$, $S1$, $S4$ and $S3$. Because there is no loop-carried dependence between $S1$ and $S4$ they can remain in the same loop, as shown in Figure 3.2(e). This can improve the locality on accesses to the b array, and eliminates the startup and iteration count overhead of one loop.

$$
\begin{aligned}
&\text{for } (i = 0; i < n; i++)\{ \\
S1: \quad &a[i] = b[i-1] + \ldots \\
S2: \quad &b[i] = c[i-1] + \ldots \\
S3: \quad &\ldots = a[i-1] + b[i] * d[i-2] \\
S4: \quad &d[i] = b[i-2] \ldots \\
&\}
\end{aligned}
$$

(a) A program with dependences.

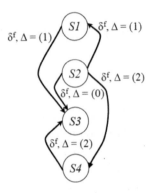

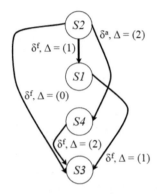

(b) The dependence graph for the program of (a).

(c) The dependence graph after it has been topologically sorted.

$$
\begin{aligned}
&\text{parfor } (i = 0; i < n; i++) \\
S2: \quad &b[i] = c[i-1] + \ldots
\end{aligned}
$$

$$
\begin{aligned}
&\text{parfor } (i = 0; i < n; i++) \\
S1: \quad &a[i] = b[i-1] + \ldots
\end{aligned}
$$

$$
\begin{aligned}
&\text{parfor } (i = 0; i < n; i++) \\
S4: \quad &d[i] = b[i-2] \ldots
\end{aligned}
$$

$$
\begin{aligned}
&\text{parfor } (i = \ldots) \\
S3: \quad &\ldots = a[i-1] + b[i] * d[i-2]
\end{aligned}
$$

$$
\begin{aligned}
&\text{parfor } (i = \ldots) \\
S2: \quad &b[i] = c[i-1] + \ldots
\end{aligned}
$$

$$
\begin{aligned}
&\text{parfor } (i = \ldots) \{ \\
S1: \quad &a[i] = b[i-1] + \ldots \\
S4: \quad &d[i] = b[i-2] \ldots \\
&\} \\
&\text{parfor } (i = \ldots) \\
S3: \quad &\ldots = a[i-1] + b[i] * d[i-2]
\end{aligned}
$$

(d) The program of (a) after being transformed to reflect the order of the topologically sorted dependence graph.

(e) The program of (a) with a statement ordering yielding slightly better locality.

Figure 3.2: An example of parallelizing loops with cross-iteration dependences.

In loops with cyclic dependence graphs with at least one loop-carried dependence, the cyclically dependent statements will form a strongly connected component (SCC) in the dependence graph. The most straightforward way to deal with the statements in each SCC is to place in a loop that is executed sequentially.

In loop nests the same strategy is followed except that it may be necessary to perform fission on all loops in the loop nest on which a pair of statements have loop-carried dependences.

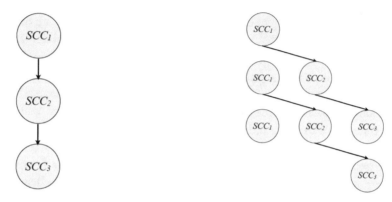

(a) A dependence graph with SCCs contracted into nodes.

(b) A pipelined execution of the SCCs across three threads.

Figure 3.3: An example of decoupled software pipelining.

In Section 3.4, we show how parallelism can sometimes be extracted from these loops using *producer-consumer* synchronization. Another way of extracting parallelism from these loops is to execute the SCCs in a pipelined fashion. This technique is called *decoupled software pipelining*, and is described in detail in [189, 190]. An example of this is shown in Figure 3.3. Note that the maximum speedup is constrained by the size of the maximum SCC. In Section 8.9, we give further readings that discuss how to overcome this impediment to performance.

3.3   TARGETING VECTOR HARDWARE

To generate code for vector hardware it is necessary to partition the statements in the original loop into a sequence of loops, each of which contains a single operation to be performed by the vector unit. The resulting loops are then further transformed to execute a sequence of vector operations on data that corresponds to the number of scalar data items operated on by each vector instruction. When performing vectorization the innermost loop will be vectorized. If a loop other than the innermost loop is more amenable to vectorization, *loop interchange* (see Section 5.5) can sometimes be used to make another loop the innermost loop.

To vectorize an inner loop, the transformation described in Section 3.2 is first applied to the loop to be vectorized, resulting in each strongly connected component in the dependence graph being in a separate loop. This yields a loop similar to those shown in 3.2(c). In the loops of Figure 3.4(a) *loop blocking* or *strip mining* has been applied to each of the resulting loops, where the resulting inner blocked loop (shown in Figure 3.4(b)) has the same number of iterations as the number of operations performed by a vector instruction. For brevity, the *fixup* code for only the last loop is shown. The fixup code is necessary since the loops being executed on the vector hardware may not have a number of iterations evenly divisible by the number of operations performed at a time by the vector hardware.

The compiler back-end generates a hardware vector instructions each of the two inner loops. One more small complexity must be handled. In Figure 3.4(b), the statement's right-hand-side (RHS) contains several operations. Vector hardware typically supports a single operation at a time (a notable exception is described below.) Either the high-level compiler that operates on loops, or the code-generating compiler, must break this RHS into multiple statements. If the statement in the strip-mined loop is viewed as a tree, as shown in Figure 3.4(c), code can be easily generated by doing an in-order traversal of the tree, and generating code for each interior node as a separate statement and loop, as shown in Figure 3.4(d).

An exception to the "one operation per instruction" rule is made for the *fused multiply add (FMA)* operation (y*z+w), which is extremely important in linear algebra. Because the multiply and add operations, and the register-to-register transfers can occur faster than a clock cycle on most processors, the FMA instruction can stream the result of the multiply operation in (y*z+w) to an adder, which takes w as the second input, returning the result of the multiply and add operations in a single cycle, giving roughly double the peak performance of performing the multiply operation and add as discrete clocked operations.

3.4   PARALLELIZING LOOPS USING PRODUCER/CONSUMER SYNCHRONIZATION

Dependences prevent parallelization because there is no guarantee the order that statements execute on different threads in the parallel version of the program will enforce the dependence, as discussed in Section 1.1. By using producer/consumer synchronization, this ordering can be forced on the program execution, allowing parallelism to be extracted from loops with dependences.

3.4.1   PRODUCER AND CONSUMER SYNCHRONIZATION

Several forms of producer/consumer synchronization (referred to simply as synchronization for the rest of this section) were implemented in the 1980s and early 1990s, during the age of the super mini-computers. Full/empty synchronization, implemented in the Denelcor HEP [126], attaches a full-empty bit to each word of storage. By having accesses either set, or reset, the full-empty bit, reads and writes to a memory word could be sequenced. Memory operations could either wait until a word was full before proceeding, and could either clear, or set, the full/empty bit.

for (i = 0; i < n/opCnt; i+ = opCnt)
   for (ii = 0; ii < opCnt; ii ++)
S4 :   d[ii] = +b[ii + 1]...

for (i = 0; i < n/opCnt; i+ = opCnt)
   for (ii = 0; ii < opCnt; ii ++)
S2 :   b[ii] = c[ii − 1] + ...

for (i = 0; i < n/opCnt; i+ = opCnt)
   for (ii = 0; ii < opCnt; ii ++)
S1 :   a[ii] = b[ii − 1] + ...

for (i = 0; i < n/opCnt; i+ = opCnt)
   for (ii = 0; ii < opCnt; ii ++)
S3 :   ... = a[ii − 1] + b[ii] ∗ d[ii − 2]
   for (i = n/opCnt ∗ opCnt; i < n; i ++)
S3 :   ... = a[i − 1] + b[i] ∗ d[i − 2]

(a) A blocked vectorized loop. All of the loop would require fixup code, for brevity it is only shown for the S3 loop.

for (i = 0; i < n/opCnt; i+ = opCnt)
   for (ii = 0; ii < opCnt; ii ++)
S5 :   d[ii] = b[ii] ∗ b[ii] + c[ii] ∗ e[ii];

(b) An example of a statement with multiple operations required by the right-hand side.

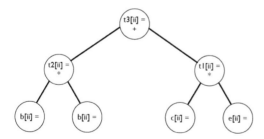

(c) An expression tree for the complex statement of (b).

for (i = 0; i < n/opCnt; i+ = opCnt)
   for (ii = 0; ii < min(opCnt, n); ii ++)
S5$_a$ t1[ii] = b[ii] ∗ b[ii];
   for (ii = 0; ii < min(opCnt, n); ii ++)
S5$_b$ t2[ii] = c[ii] ∗ e[ii];
   for (ii = 0; ii < min(opCnt, n); ii ++)
S5$_c$ t3[ii] = t1[ii] + t2[ii];

(d) Vectorized code for the statement of (b).

Figure 3.4: An example of vectorization.

The Alliant F/X 8 [1, 228] implemented the advance(r, i) and await(r, i) synchronization instructions. The advance instruction sets synchronization register r to i, and the await instruction would wait for r to take the value i. Thus, the dependence on a[i] and a[i − 1] in Figure 3.5 could be synchronized by executing advance(r, i) after the write to a[i], and await(r, i − 1) before the read. Variables guarded by the synchronization instructions could also be specified, allowing the compiler to move unsynchronized memory operations past the synchronization instructions.

Compiler exploitation of both of these synchronization instructions, and general producer/consumer synchronization, can be discussed in terms of post and wait synchronization. The instruction post(r, i, *vars*) writes the value i to r, and the instruction wait(r, i, *vars*) waits until the value of r is i. In both cases, it is assumed that no accesses on the memory locations denoted by the variables in the list *vars* are moved past the synchronization instruction by later compiler passes. The post and wait instructions also have a functionality equivalent to a *fence* instruction that will ensure that the result of all memory accesses before the post and wait are visible before the post or wait completes, and that the hardware does not move instructions past the synchronization operation at run time.

A compiler can synchronize a program directly from the dependence graph. After the source of dependence δ, it inserts the instruction post(r_δ, i, vars), where i is the loop index variable, r_δ is the synchronization register used for dependence δ, and vars contains the variables involved whose dependence is being synchronized. Before each dependence sink the compiler inserts the instruction wait(r_δ, i − d_j, vars), where d_i is the distance of the dependence on the i loop. Figure 3.5(d) shows the loop after synchronization.

As shown in Figure 3.5(c), parallelism comes from overlapping the execution of the sources of the forward dependence, from $S1$ to $S3$, on a. Figure 3.6 shows a program similar to that of the program of Figure 3.5(a) except that the cross iteration dependence on b from $S2$ to $S3$ is not present, and the backward dependence on c now has a distance of two. Synchronizing this loop as described above would lead to the execution order shown in Figure 3.6(b), and not only allows the source of the forward dependence to execute in parallel, but also allows adjacent iterations whose statements are involved in the dependence cycle to execute in parallel. This is because the distance of two on the dependences means that no backward dependence exists between adjacent iterations, and therefore the adjacent iterations can execute in parallel. Loops of this form are often referred to as *doacross* loops [61, 168] in the literature.

The reasons that producer/consumer synchronization instructions are no longer supported in hardware shows the impact that technology and economics have on what is a desirable architectural. With the rise of the *killer micros* [107], commodity processors became the economically preferred choice to use as the building block for parallel systems. Specialized synchronization instructions fell out of favor because of the increased latencies required when synchronizing across the system bus between general purpose processors, and because the RISC principles of instruction set design [105, 172] favored simpler instructions from which post and wait instructions could be built, albeit at a higher run time cost. With the rise of the multicore processor, and at least for a period of time,

```
      for (i = 0; i < n; i ++){
S1 :     a[i] = ...
S2 :     b[i] = c[i − 1]...
S3 :     c[i] = b[i − 2] + a[i − 1]
      }
```

(a) A loop with cross-iteration dependences.

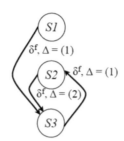

(b) The data dependence graph for (a).

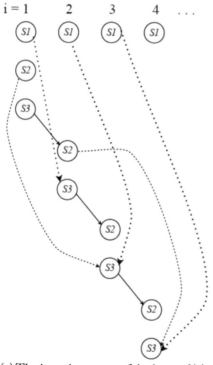

```
      for (i = 0; i < n; i ++){
S1 :     a[i] = ...
         post(0, i, a)
         wait(2, i − 1, c)
S2 :     b[i] = c[i − 1]...
         post(1, i, b)
         wait(1, i − 2, b)
         wait(0, i − 1, a)
S3 :     c[i] = b[i − 2] + a[i − 1]
         post(2, i, c)
      }
```

(c) The iteration space of the loop of (a).

(d) The loop of (a) synchronized with post/wait synchronization.

Figure 3.5: An example of using producer/consumer synchronization.

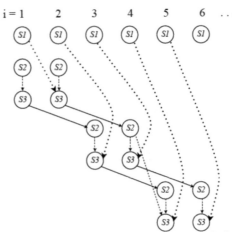

for (i = 0; i < n; i ++){
S1 :   a[i] = ...
S2 :   b[i] = c[i − 2]...
        c[i] = b[i] + a[i − 1]
    }

(a) A loop with cross-iteration dependences.

(b) The iteration space diagram of the loop of (a) showing the ordering relations imposed by the dependences.

$\delta^f, \Delta = (1)$

$\delta^f, \Delta = (2)$

$\delta^f, \Delta = (0)$

S1

S2

S3

(c) The dependence graph of the program of (a).

for (i = 0; i < n; i ++){
S1 :   a[i] = ...
        post(0, i, a)
S2 :   **wait(1, i − 2, c)**
        wait(1, i − 1, a)
S3 :   b[i] = c[i − 2]...
        c[i] = b[i] + a[i − 1]
        post(1, i, c)
    }

(d) The program of (a) synchronized.

Figure 3.6: An example of using producer/consumer synchronization for a doacross loop.

relatively low latencies in intra-processor, inter-core communication, specialized synchronization might again be attractive [145]. We note that except for questions of profitability, the compiler strategy for inserting and optimizing synchronization is indifferent to whether it is implemented in software or hardware.

3.4.2   OPTIMIZING PRODUCER/CONSUMER SYNCHRONIZATION

A compiler can sometimes reduce the number of synchronization operations needed to synchronize the dependences in a loop. While all dependences must be enforced, performing this optimization reduces the overhead of enforcing them by allowing a single post/wait pair to synchronize more than one dependence, or a combination of post/wait instructions to synchronize additional dependences. The key observation leading to the optimization is that if the dependence is enforced by a combination of program orders and other dependences' synchronization operations, no additional synchronization is needed. In the loop of Figure 3.7(a) and (b), a loop with two dependences, and the associated dependence graph, is shown.

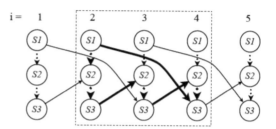

```
          for (i = 0; i < n; i++){
S1 :        a[i] = . . .
S2 :        b[i] = c[i − 1]
S3 :        c[i] = a[i − 2]
```

(a) A loop with dependences to be syn-chronized.

(b) iteration space diagram for the loop of (a). The section outlined with the dotted box is representative of a section of the ISD that is examined by the algorithm of [156] that eliminates dependences using transitive reduction.

Figure 3.7: An example of synchronization optimization.

The ISD of Figure 3.7 shows the dependences to be enforced as solid lines, and execution orders implied by the sequential execution of the program by dotted lines. Consider the dependence with distance two from statement $S1$ in iteration $i = 2$ to statement $S3$ in iteration $i = 4$. Let $S_j(k)$ represent the instance of statement S_j in iteration $i = k$. There is a path

$$S1(2) \rightarrow S2(2) \rightarrow S3(2) \rightarrow S2(3) \rightarrow S3(3) \rightarrow S2(4) \rightarrow S3(4).$$

from $S1$ in iteration 2 to $S3$ in iteration 4. If the dependence from $S3$ to $S2$ has been synchronized, then the existence of this path of enforced orders implies that the dependence from $S1(2)$ to $S3(4)$ is also enforced. Because the distances are constant, the iteration space can be covered by shifting the region in the dashed lines, and therefore every instance of the dependence within the iteration space is synchronized.

In [156] it is shown that this problem can be solved by performing a transitive reduction on the ISD. Because transitive reduction is used, it is possible for multiple dependences to work together to eliminate another dependence. The ISD on which the transitive reduction is performed needs to only contain a subset of the total iteration space (as shown by the dashed box in Figure 3.7(b)). For each loop in the loop nest over which the synchronization elimination is taking place, the number of iterations needed in the ISD for the loop is equal to the least product of the unique prime factors of the dependence distances, plus one.

Another synchronization elimination technique, described in [144], is based on pattern matching and works even if the dependence distances are not constant. The matched patterns identify dependences whose lexical relationship and distances are such that synchronizing one dependence will synchronize the other by forming a path as shown in Figure 3.7(b). In the program of Figure 3.7(a), let the forward dependences with a distance of two that is to be eliminated be δ_e, and the backward dependence of distance one be δ_1 that is used to eliminate the other dependence be δ_r. One pattern given in [144] is that if there is: (i) a path from the source of δ_e to the source of some δ_r; (ii) the sink of δ_r reaches the sink of δ_e; (iii) δ_r is lexically backward (i.e., the sink precedes the source in the program flow); (iv) the absolute value of the distance of δ_r is one; and (v) the signs of the distances of δ_r and δ_e are the same, then δ_e can be eliminated. Conditions (i) and (ii) establish the proper flow of δ_e and δ_r, (iii) recognizes that δ_r can be repeatedly executed to reach all iterations that are multiple of the distance away from the source, and (iv) and (v) say that because the absolute value of the distance is one and the signs of the two distances are equal the traversal enabled by (iii) will reach the source of δ_e.

In [157] these two techniques and others are compared and it is shown that the method of [144] actually performs better than that of [156], and has a lower compile time overhead on the loops examined.

3.5   PARALLELIZING RECURSIVE CONSTRUCTS

Iterative constructs have been the focus of most of the work on automatic parallelization for several reasons. First, in numerical programs, loops contain almost all of the work performed by the program, and as a consequence of Amdahl's law are the natural place to target for performance improvements, including parallelization. Second, in many loop based numerical programs, the data accesses tend to be amenable to dependence analysis and dependence analysis based optimizations. This in turn provides positive feedback to programmers of these applications to write code amenable to analysis. Finally, many recursive codes operate on trees or other pointer-based data structures (or equivalently, indirect accesses through arrays). These applications are fundamentally more difficult to analyze, for the reasons discussed in Section 2.1.2. Nevertheless, significant results exist for the optimization of application code containing recursion.

Among the earliest work is that of Harrison [109] on the Miprac Scheme compiler [108]. Miprac performs an abstract interpretation over a Lisp/Scheme program's recursive calls to determine if the calls can be converted to (i) a `while` loop that determines the exit condition, and (ideally) (ii) a `for` loop that can be easily parallelized and whose iteration count has been decided by the `while` loop. In some cases the `while` loop can also be parallelized (see Section 3.6).

We will briefly discuss two other approaches. Divide-and-conquer algorithms lend themselves to parallelization because at each level of recursion the new function invocations access disjoint sets of data. If a compiler can determine that this is the case it can allow each of the calls to proceed in parallel. In [90, 92], a dependence analysis similar to Eigenmann and Blume's Range Analysis [31] is used to prove the recursive calls in a quicksort operate on disjoint data. Figure 3.8(a) shows the parallelization of the quicksort algorithm using Cilk-like [51] constructs. The `spawn` constructs create two instances of the *Quicksort* function and execute then in parallel on two different threads. Because the threads operate on completely disjoint sets of data, no synchronization is needed between the sets of calls until the end of the program, where the `synch` statement ensures that all of the threads are finished before terminating the program.

```
Main( ) {
    Type sortArray[lb:ub];
    Quicksort(sortArray, lb, ub);
    synch
}

Quicksort(sortArray, low, high); {
    if (low < high) {
        q = Partition(sortArray, low, high);
        spawn Quicksort(sortArray, low, q);
        spawn Quicksort(sortArray, q+1, high);
    }
}
```

(a) Recursive routines that access independent data.

```
class dagNode {
    private static int total;
    private boolean visited;
    private dagNode leftChild, rightChild;
    void traverse(int i) {
        total = total + i;
        if (!visited) {
            visited = true;
            if (leftChild != null)
                traverse(this.leftChild, i);
            if (rightChild != null)
                traverse(this.rightChild, i);
        }
    }
}
```

(b) Recursive routines whose operations commute.

Figure 3.8: Examples of parallelizable recursive code.

Another approach targets object oriented languages and uses commutativity analysis to determine when operations performed in multiple invocations of a function are commutative [197]. When this is true, all possible interleavings of operations across parallel invocations of the function will give the same result as a sequential execution of the function invocations. To determine commutativity, two tests need to be performed.

1. Test that values of instance variables are the same whether two operations Op_1, Op_2 (which may be method calls) execute in the order Op_1 then Op_2, or the order Op_2 then Op_1. This ensures that the outcomes of the operations are the same regardless of the order they execute in. An important case where this is true is when two operations operate on independent data, or the data accessed are read only—these operations always commute.

2. The set of operations performed when executing Op_1 and then Op_2 are the same as the set of operations performed when executing Op_2 and then Op_1. This ensures that the multiset of operations that execute in either order are the same, and allows the analysis of (1) to show

that if all operations are commutative, then because the multiset of operations executed is the same in either order, then the outcomes of the complete operation must be the same.

The parallelizable invocations of the Quicksort routine are detected with commutativity analysis because the two function calls access disjoint data sets and are therefore independent, and commutative. The function calls of the traverse routine in Figure 3.8(b) are also parallelizable using commutativity analysis even though the data accessed is not independent. If a naive parallelization of the calls is performed, there is the danger of a race condition on the check of whether the function has been previously visited. By enclosing the accesses to the read and update of visited array in a mutex, the race will be avoided, and commutativity analysis inserts this synchronization.

3.6   PARALLELIZATION OF WHILE LOOPS

The parallelization of while loops is problematic for two reasons. First, unlike counted loops (e.g., a simple C for loop or Fortran do loop) the number of iterations is usually not known before the loop begins executing, making it difficult to spread the loop iterations across processors and to determine when to terminate the loop. Second, while loops are often used to perform computation on less regular data structures, including pointer based structures such as linked lists, trees, and the like, or array accesses involving non-linear subscripts, often in the form of subscripted subscripts (e.g., a[b[i]]). We will discuss the issues raised by each of these in turn.

Figure 3.9(a) shows a stylized while loop. The loop has four parts—the initialization, shown in $S1$, the test, shown $S2$, the increment, shown in $S4$, and the work or remainder, shown in $S3$ and $S5$. We now discuss two challenges in parallelizing the loop: determining the number and distribution of iterations in the loop and exploiting parallelism in both determining the loop iterations and in the work done by the body of the loop.

3.6.1   DETERMINING ITERATIONS TO BE EXECUTED BY A THREAD

The form of the update in the stereotypical while loop is shown in Figure 3.9(a). In the best situation, the loop iterations are controlled by a simple induction of the form $a \cdot i \pm c$, where a and c are loop invariants. If the test function τ (see statement $S2$) is a simple comparison (e.g., $i < n$) the while can be converted to a simple for loop. If the test is more complicated, then code such as is shown in Figure 3.9(b) can be used. The array first captures the first (lowest) iteration in each thread that meets the termination condition. After all threads complete their iterations, or meet the termination condition, the lowest iteration that met the termination condition across all threads is found. Threads that have executed iterations beyond that will need to roll back their state, as discussed in Section 3.6.2. If iterations are executed in order, then when the terminating iteration is encountered by a thread the iteration value can be sent to all threads, and they only need to complete iterations up to that value.

A variant of this form is the iteration sequence of the form of the recurrence $x(i) - a \cdot x(i - k) + c$, where a, k and c are all loop invariant, as shown in Figure 3.9(c). The loop is split into two

$S1$ ctl = init
$S2$ while(τ(ctl)){
$S3$ $work_1$(ctl, state);
$S4$ update(ctl, state);
$S5$ $work_2$(ctl, state);
        }

(a) a stereotypical `while` loop.

SO first[:] = maxint;
$S1$ ctl = init
$S2$ parallel for (i = 0; i < maxint; i ++){
$S2_a$ if (!τ(ctl) first[threadID] = i {
$S2_b$ first[threadID] = i)
$S2_c$ first[threadID] = i)
        } else {
$S3$ $work_1$(ctl, state);
$S4$ update(ctl, state);
$S5$ $work_2$(ctl, state);
        }
    }

(b) A parallel execution of the loop where iterations are an induction sequence.

$S1$ ctl = 1
$S2$ while(τ(ctl) < u){
$S3$ $work_1$(ctl, state);
$S4$ ctl = a · ctl + c;
$S5$ $work_2$(ctl, state);
        }

(c) A loop whose iteration sequence is a recurrence.

$S1$ i = 2
$S1_a$ ctl[1] = 1
$S2$ while(τ(ctl[i − 1]) < u){
$S4$ ctl = a · ctl[i − 1] + c;
$S4_a$ i = i + 1;
        }
$S4_b$ N = i

$S2$ parfor (i = 0; i < N; i ++)
$S3$ $work_1$(ctl[i], state);
$S5$ $work_2$(ctl[i], state);

(d) The loop parallelized. Note that the first loop can be done in parallel using parallel prefix.

Figure 3.9: Examples of some issues in parallelizing `while` loops.

parts – the first solves the recurrence and the second does the actual work in the loop. The recurrence can be further optimized by being solved in parallel using a parallel prefix algorithm that finds the necessary partial results of the recurrence in time proportional to the log of the number of iterations in the loop.

More complicated iteration sequences often require a sequential computation. An example of this is a `while` loop that iterates over a linked list, with the `work` function operating on each element of the list. In [193], three ways of dealing with this problem are outlined, as shown in Figure 3.10. Figure 3.10(a) shows the original loop, and the other parts of the figure show three strategies for parallelizing the loop.

Figure 3.10(b) shows a parallel version of this loop. A global (gP) and local (lP) pointer into the linked list are maintained. When a process needs another work item, it atomically assigns what is pointed to by the global pointer to the local pointer, moves the global pointer to the next element in the linked list, and then works on the node pointed to by the local pointer. The overhead of synchronization, and the coarse grained synchronization (i.e., locking the entire list) limit the parallelism of this approach. We note that more intelligent data structures and a compiler that is aware of their semantics [136, 137, 236] might at least partially ameliorate some of these drawbacks.

If there are $|T|$ threads, the technique demonstrated in Figure 3.10(c) will have thread t_i work on nodes $i, i + |T|, i + 2|T|, \ldots$ in the linked list. Each thread visits all of the nodes, but only performs work on those that is supposed to work on. Thus, the loop at $S1_a$ executing in thread t_i skips over the first $i - 1$ nodes to find the first node to work on. Afterwards, the loop at $S1_e$ skips over the $|T|$ nodes worked on by the other threads to find the next node to work on. This technique eliminates the synchronization overhead of the first method, but each process will need to look at all N nodes, and so the amount of work done by each processor has an $O(N)$ term, which will necessarily limit the achievable speedup.

The third method (Figure 3.10(d)) has the same overheads as the second, but can achieve a better load balance by virtue of assigning nodes to threads as they need work, rather than having an implicit pre-assignment of nodes as seen in the second method of Figure 3.10(c). As each thread gets an iteration i to execute, it examines the previous iteration (stored in prevIter) it executed. The difference between the current and previous iteration is the number of nodes processed by other threads, and these nodes are skipped over in the loop beginning with statement $S1_c$.

3.6.2 DEALING WITH THE EFFECTS OF SPECULATION

In the above examples it was known that a linked list was being accessed, and therefore it is also known that a traversal of the list will visit unique nodes. Depending on other data pointed to by these nodes (and accessed in the work routine) dependences may exist at run time among concurrently executing invocations of the work routine. Moreover, it is sometimes more efficient to speculatively execute iterations of a while loop under the assumption that the loop will not terminate before those iterations are executed. This assumption will not always hold, and in those cases it will be necessary to ensure that the effects of those excess iterations are not reflected in the permanent program state.

Two approaches can be used to solve this problem. First, techniques like transactional memory [101, 158] can be used to buffer the state until it is safe to commit. Unfortunately, software transactional memory carries significant performance penalties, and hardware transactional memory is not yet available on widely deployed processors. A second approach is to checkpoint the state accessed within a loop, and if the loop fails in some way, to restore the checkpoint and execute the loop sequentially. By *strip mining* the loop (see Section 5.1), its execution can be broken into blocks of contiguous iterations. A checkpoint is taken for each block, and when a block successfully finishes its values are committed, its checkpoint is discarded, and a new checkpoint is taken for the next

```
S1    Node ∗ p = head(list);
S2    while (i = 1, u){
S3        work(p);
S4            p = next(p);
          }
```

(a) A `while` loop with pointer-based data structures.

```
S1    Node ∗ gP = head(list);
S2    parfor (i = 1, u) {
S1a       Node ∗ lP = head(list);
S1b       lock(list);
S1c       gP = next(gp);
S1d       unlock(list);
S3        work(p);
          }
```

(b) One possible parallelization of the while loop.

```
S2    parfor (i = 1, |T|) {
S1a       Node ∗ lP = head(list);
S1b       for (int j = 0, j < threadID; j++) {
S1c           lP = next(lP);
S1d           if (lP == null) goto EXIT;
          {
S3        work(p);
S1e       for (int j = 0, j < |T|; j++){
S1f           lP = next(lP);
S1g           if (lP == null) goto EXIT;
S1h       if (lP == null) goto S3;
          }
      EXIT:
```

(c) A parallelization that uses a fixed assignment of nodes to processors and no synchronization.

```
S2    parfor (i = 1, u) {
S1a       Node ∗ lP = head(list);
S1b       int prevIter = 0;
S1c       for (intj = 1, j < i − prevIter; j++) {
S1d           lP = next(lP);
S1e           if (lP == null) goto EXIT;
          }
S3        work(p);
S3        prevIter = i;
          }
      EXIT:
```

(d) A parallelization that does uses a dynamic assignment of nodes to processors and no synchronization.

Figure 3.10: An example of parallelizing a `while` loop traversing a pointer-based data structure.

block. This both reduces the memory footprint required and prevents a failure late in the execution of the loop from discarding all work that has been done so far in the execution of the loop.

Additional readings on more advanced speculative techniques can be found in Section 8.9.

3.7 SOFTWARE PIPELINING FOR INSTRUCTION LEVEL PARALLELISM

Even after applying the parallelization techniques above, additional instruction level parallelism can be exposed in loops using software pipelining, which can result in significant additional speedups. Software pipelining rearranges instructions across loop iterations to expose the parallelism to the instruction scheduler within a processor, or to exploit VLIW [69, 81] processors. Consider the loop of Figure 3.11(a) and (b). Assume it takes one cycle to address a[i], to load an element of a, to perform a multiply operation, to store the result, to check the value of the index and increment it, and one cycle to perform the branch the branch. Under these assumptions, one iteration of the loop will take six cycles, and computing and storing n values will take $6n$ cycles. In the code in Figure 3.11(b), the timing assumptions are the same as before, but the hardware will execute multiple statements on a single line in one cycle. The transformed loop uses instruction level parallelism to execute an iteration of the loop in five cycles, but each iteration produces two results, for an average of 2.5 cycles a result, or $2.5n$ cycles total, ignoring the startup code.

```
float k, a[n];
for i = 0; i < n; i++
    a[i] = a[i] * k
}


r7 = &a[0]
r8 = &a[n]
r5 = k;
L: {
    r1 = (r7);
    r1 = r1 + k;
    (r1++) = r1 + k;
    r4 = &a[i + 5]; r11 = r7 < r8?
    if r11, goto L
}
```

(a) A simple `for` loop and assembly code that might be generated for it.

```
S1   r7 = &a[0]
S2   r8 = &a[n]
S3   r5 = k;
S4   r1 = &a[0];
S5   r1 = r1 * r5;
S6   r2 = &a[1];
S7   r2 = r2 * r5;
S8   r3 = &a[2];
S9   r4 = &a[3];
S10 L: {
S11     (r7) = r1; r1 = (r3) * (r5)
S12     (r7, 4) = r2; r2 = (r4) * (r5)
S13     r3 = (r7, 16); r10 = (r7, 8)
S14     r4 = (r7, 20); r11 = r10 < r8?
S15     r7+ = 8; if r11, gotoL
     }
```

(b) Assembly code for the software pipelined version of the loop in (a).

Figure 3.11: An example of software pipelining.

Software pipelining, like all transformations that alter the order of memory fetches, must ensure that the new access order enforces all dependences. In statement $S11$ of Figure 3.11(b), the value of a stored in some iteration i is occurring in parallel with the fetch and add of values from iteration a[i + 2]. In statement $S12$ a store of a value of a from iteration i + 1 is occurring in parallel with the fetch and add of a that would normally be done in iteration i + 3. As well, in statements $S13$ and $S14$ performs fetches that would normally be done in iterations i + 4 and iteration i + 5. Depending on the actual subscripts of a, these fetches and stores could be violating dependences, and would prevent software pipelining.

CHAPTER 4

Transformations to modify and eliminate dependences

If an original, unchanged loop is analyzed, dependences will almost certainly be found that prevent useful transformations, such as tiling, interchange and parallelization, from being performed. In this chapter, several transformations that modify, or completely eliminate dependences will be described. These transformations typically do not lead to an increase in performance—on the contrary, they may lead to more loop bounds and subscript index expressions and computation, resulting in more work being performed. They can, however, enable transformations that dramatically increases the performance of the program.

There is another class of transformations—reduction and recurrence recognition—that allow dependences to be ignored because the operations on the data involved in the dependence are commutative, but do not actually remove them. We discuss reductions in Sections 4.8 and give pointers to readings on recurrence recognition in Section 8.5.

4.1 LOOP PEELING AND SPLITTING

Loop peeling [248] is a transformation that removes a small number of iterations from the loop, reproducing them as straight-line code before or after the loop. Loop splitting is essentially the same transformation except that enough iterations are peeled off of the loop that executing them in a second loop is the best way to generate code.

In the loop of Figure 4.1(a), the last value for the array a is captured in the variable last. Standard dependence testing will assume that there are loop carried anti, output and flow dependences on last. By peeling off the last iteration of the loop, the remaining iterations have no loop carried dependences and therefore the loop can be executed in parallel.

In the loop of Figure 4.1(c), there are no loop-carried dependences within the first half of the iteration space, and there are none within the second half of the iteration space, but there are loop-carried dependences from the first half to the second half of the iteration space. The table of Figure 4.1(d) shows the elements of a accessed on each iteration, and the dependence relation between the two halves of the iteration space can be quickly inferred from the references. By using loop splitting to divide the loop into two loops, one running from $0 \ldots n/2 - 1$ and the other from $n/2 \ldots n - 1$, two loops are formed that have no dependences and can therefore be made fully parallel.

Because the order of access is not changed for any reference, the transformation is always legal. The only difficulty for the compiler is determining when to profitably apply the transformation. This is generally done in a somewhat ad-hoc manner by having the compiler look at suspicious dependences to see if, e.g., a scalar references involved in loop-carried dependences is control dependent on an condition such that the scalar is only accessed in a single iteration.

```
for (i = 0; i < n; i ++) {
    float a[n];
    t[i] = ...;
    a[i] = f(t[i], ...);
    if (i == n − 1) last = a[i];
}
```

(a) A loop with a dependence causing scalar access in the last iteration.

```
{
    float a[n];
    for (i = 0; i < n; i ++) {
        t[i] = ...;
        a[i] = f(t[i], ...);
    }
    last = a[n − 1];
}
```

(b) The loop of (a) with the last iteration peeled off, allowing the loop to be parallelized.

```
for (i = 0; i < n; i ++){
    a[i] = a[n − i]
}
```

i	a[i]	a[n-i]
0	**0**	9
1	**1**	8
2	**2**	7
3	**3**	6
4	**4**	5
5	5	4
6	4	3
7	3	2
8	2	1
9	1	0

(c) A loop where dependences on a exist between the first half of the iteration space to the last half.

(d) The elements of a accessed in each iteration, showing that the iterations in each half of the iteration space are dependence free. The first half of the iteration space is shown in bold.

Figure 4.1: An example of loop peeling and loop splitting.

4.2   LOOP SKEWING

Loop skewing changes the direction vector of a dependence, enabling loop parallelization and interchange (see Section 5.5). Consider Figure 4.2(a) and (b), which show a loop and its associated iteration space diagram. The dependence on a has a direction of $(<, >)$, which precludes parallelization on both loops and prevents interchanging the i_1 and i_2 loops for better cache locality.

Increasing the sink iteration of the i_2 loop by two iterations would give a direction of $(<, =)$, and by more than two iterations would give a "$<$" direction. The case where the sink iteration of the i_2 increases by two is shown in Figure 4.2.

Let i_{out} be an outer loop, and i_{in} some inner loop. Let δ be a dependence whose source and sink iterations for some loop i are i_{src} and i_{sink}, respectively. The distance of the dependence is $D = [\dots, d_{i_{out}}, \dots, d_{i_{in}}]$. Skewing allows us to shift the sink iteration of the i_{in} dependence by the amount

$$k \cdot d_{i_{out}}, k > 0$$

where $k = 0$ is the case when skewing is not done.

The transform itself is straightforward. The shifting of iterations is achieved by altering the i_{in} loop bounds by adding i_{out} to both the lower and upper bounds. Where previously the lower bound of the i_{in} loop was $L_{i_{in}}$, the new lower bound at the dependence source is $L_{i_{in}} + i_{out,src}$ and the new lower bound at the dependence sink is $L_{i_{in}} + i_{out,src}$. This shifts the iteration space of the instance of the i_{in} loop at the dependence source by $i_{out,src} - i_{out,sink}$ iterations, i.e., by $d_{i_{src}}$ iterations. Because the values taken on by the loop index variables values have changed, the values of the subscript expressions must also change. By changing each i_2 term in the subscript expression to $i_{in} - i_{out}$, the subscript expression values will be the same as in the original loop.

If the original resulting dependences still have an undesired $>$ direction in the i_{in} element of the loop, the above transformation can be repeated, shifting the lower bounds of the i_{in} loop by $L_{i_{in}} + 2 \cdot i_{out,src}, L_{i_{in}} + 3 \cdot i_{out,src}, \dots, k \cdot i_{out}$, and so forth. In practice, given distance $d_{i_{out}}$ and $d_{i_{in}}$,

$$k = \left\lceil \frac{|d_{i_{in}}|}{|d_{i_{out}}|} \right\rceil$$

is used to perform the skewing.

We note that skewing does not change the order in which dependence source and sinks execute, it only changes the "labels", i.e., the values, of those iterations. Thus, skewing is always legal to perform.

Loop normalization is a form of skewing (or, more precisely, deskewing [250]) that is often used to make it easier to form dependence equations [244]. Loop normalization transforms the loop in Figure 4.3(a) into the loop of Figure 4.3(b). Since loop normalization is a form of skewing, it too is always legal to perform. As shown in the figure, normalization takes the loop from $L_i \leq i < U_i$, and converts it to a loop starting at zero, or one, i.e., to a loop from $0 \leq i < U_i - L_i$. If $L_i > 0$,

```
for (i₁ = 1, i₁ < n; i ++) {
    for (i₂ = 1, i₂ < n; i₂ ++) {
        a[i₁, i₂] = a[i₁ − 1, i₂ + 2]
    }
}
```

(a) A loop with a $(<, >)$ dependence.

```
for (i₁ = 1, i₁ < n; i₁ ++) {
    for (i₂ = k * i + 1,
         i₂ < k * i₁ + n; i₁ ++) {
        a[i₁, i₂ − k * i₁] =
            a[i₁ − 1, i₂ − k * i₁ + 2]
    }
}
```

(b) The loop skewed by k times the distance of dependences on the i_1 loop.

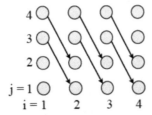

(c) The iteration space diagram for the loop of (a).

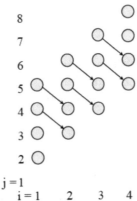

(d) The iteration space diagram for the loop of (b) when the skewing factor $k = 1$.

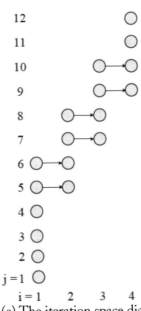

(e) The iteration space diagram for the loop of (b) when the skewing factor $k = 2$.

Figure 4.2: An example of loop skewing.

which it typically is if it is not equal zero. Thus, normalization runs the risk of making an "=" or "<" dependence into a ">" dependence.

$$\text{for } (i = L_i, i \leq U_i; i+ = s) \{ \qquad \qquad \text{for } (i = 0, i \leq U_i - L_i; i+ = 1) \{$$
$$\dots a[f(i)] \dots \qquad \qquad \qquad \dots a[f((\cdot s + L_i)] \dots$$
$$\} \qquad \qquad \qquad \qquad \qquad \}$$

(a) The loop before normalization.   (b) The loop after normalization.

Figure 4.3: An example of loop normalization.

4.3   INDUCTION VARIABLE SUBSTITUTION

A common programming idiom is to have an *induction* variable that is, implicitly or explicitly, a function of the loop induction variable. An example of this is the concatenation of one vector onto another, as seen in Figure 4.4(a). When attempting to parallelize the loop, the references to k are problematic. In particular, every iteration of the loop reads and writes the same memory location, causing a cross-iteration dependences between the read of k in $S2$ and the write of k in $S3$.

It is easily observed that at some iteration i = v, the value of k in $S2$ will be start + i. Recognizing this, and making the proper substitution into the use of i in $S2$ enables the elimination of statement $S3$. This in turn eliminates the cross-iteration anti, output and flow dependences on k. The loop after this transformation is shown in Figure 4.4(b).

More complicated situations can also be handled by induction variables substitution. In Figures 4.4(c) and (d), induction variable substitution is applied to triangular loops, and Figures 4.4(e) and (f) show it applied to multiple loop index variables.

Induction recognition proceeds as follows. First, all variables whose value is a function of only index variables, constants, and, recursively, other variables with the same properties are identified. The form of the expression is analyzed and a closed form expression in terms of loop index variables is generated. Next, the closed form expression is either assigned to a private variable that is substituted into all uses of the variable, or the expression itself is substituted into uses of the induction variable. The process continues until all induction variables have been expressed in terms of constants and loop index variables.

4.4   FORWARD SUBSTITUTION

Programmers often assign expressions into scalar temporaries to be reused throughout a computation. Compilers often perform the same optimization (called *common sub-expression elimination*) where repeatedly used expressions, or parts of an expression, are identified, with the value of the expression assigned into a temporary. If this is done within the body of the loop, a cross-iteration output dependence will exist on the scalar temporary. An example of this is shown in Figure 4.5.

```
S1   k = start;
     for (i = 0; i < n; i ++) {
S2       a[k] = b[i];
S3       k ++;
     }
```

```
     for (i = 0; i < n; i ++) {
S2       a[start + i] = b[i];
     }
```

(a) k is a candidate for induction variable substitution.

(b) The loop of (a) after induction variable substitution.

```
S1   k = start;
     for (i1 = 0; i1 < n; i1 ++) {
       for (i2 = 0; i2 < i; i2 ++) {
S2         k ++;
S3         a[k] = b[i];
       }
     }
```

```
S1   k = start;
     for (i1 = 0; i1 < n; i1 ++) {
       for (i2 = 0; i2 < i; i2 ++) {
S2         a[i1 * n + i2 + start + 1] = b[i1];
       }
     }
```

(c) k is a candidate for induction variable substitution within a triangular loop nest.

(d) The loop of (c) after induction variable substitution.

```
S1   k1 = 0; k2 = 0;
     for (i1 = 0; i1 < n; i1 ++) {
       for (i2 = 0; i2 < i1; i2 ++) {
S2         k1 = k1 + 1;
S3         a[k2] = ...;
       }
S4       k2 = k2 + k1;
     }
```

```
S1   k = start;
     for (i1 = 0; i1 < n; i1 ++) {
       for (i2 = 0; i2 < i; i2 ++) {
S2         a[k2private] = ...;
       }
S4       k2private =
             ((i1 * i1 * i1 + 2 * i1)/3 - i1)/2;
     }
```

(e) A loop with coupled induction variables in a triangular loop.

(f) The loop of (c) after induction variable substitution.

Figure 4.4: Examples of induction variable substitution.

Let *tmp* be the problematic temporary. Let *expr* be the right-hand side expression, and let V_{expr} be the variables read in *expr*. Assume that no value of any $v \in V_{expr}$ (or any variable aliased to v) is changed between the assignment to *tmp* and a use of *tmp*. Then *expr* can be substituted into every use of *tmp*, eliminating all uses of, and the stores into, *tmp*, and therefore all dependences involving *tmp*. This transformation does increase the amount of work done in the loop, and in loops with a great deal of integer arithmetic (e.g., array subscript expressions) this integer arithmetic can form the critical path through the loop. Thus, forward substitution should be used when profitable, i.e., when it enables another transformation.

```
          for (i = 0; i < n; i ++) {
S1            tmp = expr;
S2            ... = a[i + tmp − 1];
S3            a[i + t₁] = a[i + tmp − 1];
          }
```

```
          for (i = 0; i < n; i ++) {
S2            ... = a[i + expr − 1];
S3            a[i + t₁] = a[i + expr − 1];
          }
```

(a) A program that uses a temporary *tmp* to store a reused expression value.

(b) The program of (a) after forward substitution.

Figure 4.5: An example of forward substitution.

4.5   SCALAR EXPANSION AND PRIVATIZATION

Scalar expansion and privatization attempt to eliminate the same sort of dependences that forward substitution does, but do so by using additional storage rather than additional computation. The basic idea is that if (i) a scalar's value is always computed in an iteration of the loop before it is used, and (ii) the value read within an iteration is always the value computed in that iteration, then each iteration's value can be stored in separate memory location. Since each iteration reads and writes from a separate location, there are no loop-carried dependences on the variable.

Figure 4.6(a) shows a loop amenable to this transformation, with t being the scalar to be expanded or privatized. Scalar expansion is legal when the following conditions are true:

1. No use of the variable is *upward exposed*, i.e., the use never reads a value that was assigned outside the loop.

2. No use of the variable is from an assignment in an earlier iteration.

Whether or not these conditions are met can be determined by consulting the use-def information for the variable (see Section 2.5.) For the loop of Figure 4.6(a) the conditions are met. Making the data read and written to t private to an iteration can be accomplished in two ways, as we now describe.

Figure 4.6(b) shows the first way of doing this by *expanding* a scalar into an array of elements, with one element per loop iteration. Since all writes and reads of a given element of the array are within a single iteration, there are no loop-carried dependences on the array. We observe that (despite what we have said earlier when discussing parallelization) the real problem is not dependences that cross just any iteration, but rather dependences that go from an iteration executing on one thread to an iteration executing on another thread. Because the iterations on a thread execute in-order, dependences spanning iterations within a thread are enforced by the sequential execution of the program within the thread. This observation can be used to generate the code found in Figure 4.6(c), where storage for a *private* scalar is allocated on a thread local stack, as denoted by the `private` keyword. This solution is preferable to scalar expansion as it does away with the computational overhead of address calculation when using an array, reduces the storage overhead to be on the order of the number of threads rather than on the order of the number of iterations, and increases locality.

Privatization also decreases the cache footprint of the t array and eliminates false sharing across cores and threads.

```
                                         for (i = 0; i < n; i ++) {
    for (i = 0; i < n; i ++) {      S0      int t[n];
S1      t = e;                      S1      t[n] = e;
S2      a[i + t] = a[i + t − 1];    S2      a[i + t[n]] = a[i + t[n] − 1];
    }                                    }
```

(a) A temporary t amenable to scalar expansion.

(b) The loop after performing scalar expansion.

```
                    for (i = 0; i < n; i ++) {
            S0          private int t;
            S1          t = e;
            S2          a[i + t] = a[i + t − 1];
                    }
```

(c) The loop after performing privatization.

Figure 4.6: An example of scalar expansion and privatization.

4.6   ARRAY PRIVATIZATION

Like scalars, arrays also are sometimes used as temporaries within a loop nest and can be profitably parallelized. An example of this is shown in Figure 4.7(a) and its dependence graph in Figure 4.7(b).

The problem with the program in the example, and in general, is that elements of a d-dimensional array are repeatedly assigned in a $d + k$ deep loop nest, where $k > 0$. This creates loop carried dependences on all the loops in the loop nest. Note that scalar privatization, discussed above, is the degenerate case where $d = 0$. An additional difficulty is that different iterations may assign the final value into different elements of the array, and if these values are used outside of the loop nest they need to be assigned into the shared, global version of the array.

To perform privatization on some array a to eliminate loop carried dependences on loop i the analysis must first determine that all values read from the array in some iteration of loop i are also written in that iteration. One way to determine this is to determine the sections of the array that *must* be written in an iteration prior to each read of the array. If it is true that all elements read from the array at each read in the iteration must have also been written earlier in that iteration, then it is legal to perform privatization. Because the analysis is checking for prior reads in the same iteration, the analysis for legality does not need to worry about accesses in other iterations. Thus, a single forward pass through the loop body can be made to gather the data about reads and writes.

Data that is read or written by some reference r can be represented by a *regular section descriptor* $[L_r : U_r : s]$, where L_r is the first element of the array that is accessed by r, U_r is the last element

```
double a[n];
...
for (i₁ = 1; i₁ < n; i₁ ++) {
S1      a[i₁] = ...
        for (i₂ = 2; i₂ < i₁ + 1; i₂ ++) {
S3          a[i₂] = ... + a[i₂ − 1];
        }
        for (i₃ = 2; i₃ < i₁; i₃ ++) {
            b[i₁, i₃] = ... + a[i₃ + 1];
        }
}
... = a[σ];
```

(a) A program amenable to array privatization.

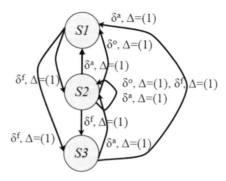

(b) The dependence graph for the program.

```
double a[n];
...
for (i₁ = 1; i₁ < n; i₁ ++) {
    private double a′[n];
    a′[i₁] = ...
    for (i₂ = 2; i₂ < i₁ + 1; i₂ ++) {
        a′[i₂] = ... + a′[i₂ − 1];
    }
    for (i₃ = 2; i₃ < i₁; i₃ ++) {
        b[i₁, i₃] = ... + a′[i₃ + 1];
    }
}
... = a[σ];
```

(c) The program after privatization but without last writes being performed on the original a array.

```
double a[n];
...
for (i₁ = 1; i₁ < n; i₁ ++) {
    private double a′[n];
    a′[i₁] = ...
    for (i₂ = 2; i₂ < i₁ + 1; i₂ ++) {
        a′[i₂] = ... + a′[i₂ − 1];
    }
    for (i₃ = 2; i₃ < i₁; i₃ ++) {
        b[i₁, i₃] = ... + a′[i₃ + 1];
    }
    if (i₁ = n − 1) a[1 : n] = a′[1 : n];
}
... = a[σ];
```

(d) The program after privatization but without last writes being performed on the original a array.

Figure 4.7: An example of array privatization.

accessed, and s is the stride, or distance between adjacent elements accessed by r. A non-unit stride can result from either a non-unit coefficient of an index variable, or a non-unit stride for a loop index variable. Sections are able to precisely represent the region of an array accessed by a counted `for` loop with affine subscript expressions (see Section 2.3.2), i.e., subscripts that are linear expressions in one or more dimensional spaces. Thus, in $S1$ above, the elements written in a single iteration of the i_1 loop would be $[i_1 : i_1 : 1]$, and in $S2$, the elements read in a single iteration of the i_2 loop would be $[2 : i_1 : 1]$.

Remember that for array privatization to be performed it is necessary that all values read from an array in an iteration *must* be written earlier in that iteration. Thus, the elements defined (i.e., written) in an iteration of the loop in which the array is being privatized must be a (not necessarily proper) superset of the elements read. Therefore, under-estimating the set of elements written by a reference will lead to lost opportunities for privatization, but not to an incorrect privatization. Similarly, over-estimating the set of elements read is also a conservative approximation of the actual elements read.

This conservative approach allows an accurate dataflow analysis to be developed to determine the legality of privatization. In particular, a single forward pass is made to determine what elements of the array have been written in an iteration. At the join of two control flow paths through the loop (typically because of an if statement) the sections describing the elements that have been written along each path are intersected, leading to a possible under-estimation of the elements defined. Along a path, the sections corresponding to the elements defined by multiple assignments to the array can be formed by unioning the sections if they overlap. Then, at each read, a check is made to see if the set of array elements that have been defined up to that point in the iteration is at least as large as what is being read. If this is true at all reads (uses) of the array in the loop, the array can be legally privatized.

The issue of storing last writes to an external array must also be handled if the array privatization is to be legal. In the example program, the writes to a in both $S1$ and $S2$ store into $a[i_1]$. The write in the i_2 loop is the second, and last write of each iteration, but later iterations may also write the array. In some cases, such as that shown in the example, it can be determined that a particular write to the array being privatized will be the last write. Because the i_2 loop writes increasingly large portions of the array that subsume the elements of the array written in the last iteration of the outer i_1 loop, it can be determined that the writes to a in the i_2 loop when $i_1 = n - 1$ will be the last writes into the element. More generally, a data structure can be kept that records the iteration that the last write of each element occurred. When the current iteration is greater or equal to this count, the global array is updated with the value being written to the privatized version of the array. The "equal" part of the test is necessary when multiple writes to an element occur in the same iteration of the loop nest. Again, in this case compiler analysis can often determine which will be the last write in an iteration, and eliminate the test for the earlier writes. This is particularly easy when there are no zero-trip loops or if branches containing writes to the privatized array.

Privatization is considered profitable if it eliminates dependences. This occurs whenever it is legal and the array being privatized has one or more elements read in different loop iterations.

4.7   NODE SPLITTING

Dependence cycles involving both flow and anti dependences can often be broken with the *node splitting* transformation. This transformation is enabled by the fact that anti dependences arise from the re-use of storage, i.e., the read of a memory location is followed by a write to the same location.

By adding extra storage to save the read location and ensure it is not overwritten, the location of the read can be moved about the loop, thereby breaking the cycle.

Figure 4.8 gives an example of this. As can be seen in Figure 4.8(a), there is a lexically forward flow dependence from $S1$ (a[i]) to $S2$ (a[i-1]), and a lexically backward anti-dependence from $S2$ (a[i+1]) to $S1$ (a[i]), leading to the cycle shown in the dependence graph. The transformation identifies cross iteration anti-dependences, and assigns the value at the source of the dependence (a[i+1], in this case) to a temporary variable. As shown in Figure 4.8(b), this is insufficient to break the cycle. Now, because the anti-dependence is loop-carried, the assignment of the source to a temporary variable can be moved within the body of the loop, i.e., within a particular iteration, without changing the value read. By moving the statement before the use of the temporary variable (i.e., the original source of the anti-dependence) the lexically backward anti-dependence becomes lexically forward, potentially breaking the cycle. This is shown in Figure 4.8(c). Moving the assignment as early as possible in the loop is legal, but leaving it as lexically forward in the loop as it can be while still making the anti-dependence lexically forward may reduce the time that the value of the temporary is stored in a register, and increase the effectiveness of the register allocator.

While both privatization and node splitting target anti dependences, they are used in different situations. Privatization is useful when a variable is assigned a value that is only used within that iteration. Thus, the potential anti and output dependences do not really exist (in terms of the flow of data values in a sequential execution of the program), but parallelizing the loop and having multiple reads and stores, in an indeterminate order, to the single storage location will introduce races and be incorrect. Node splitting is useful when the anti-dependence may actually exist at run time, and thus the global state of the involved variables must be maintained. Because the two transformations are useful in different situations, they are complementary and can be used together.

4.8   REDUCTION RECOGNITION

The last technique we will discuss to remove dependences is *reduction recognition*. Reduction recognition exploits associativity to both break dependences and enable code motion and parallelization. Reductions are operations that reduce the dimensionality of at least one input operation using a commutative *reduction operation*, $\oplus$. Statement $S2$ in Figure 4.9(a) shows a reduction which reduces the one-dimensional array a into the scalar s where the reduction operation is $+$. This follows the general form of a reduction of $s = s \oplus a[\ldots]$

Figure 4.9(b) shows the parallelization of the reduction in Figure 4.9(a). The transformation creates a private storage location for each thread involved in the parallel execution of the loop. Each thread then computes the partial sum involving the elements of the reduced variable (a in this case) accessed in the iterations it executes. A second loop sums the partial sums to give the final value of s.

A better parallelization of the reduction is shown in Figure 4.9(c). Using $\log_2(|T|)$ steps, where $|T|$ is the number of threads, the summation of the partial sums is parallelized, as shown in

```
            for (i = 0; i < n; i ++) {
      S1      a[i] = . . . ;
      S2      . . . = a[i + 1] + a[i − 1];
            }
```

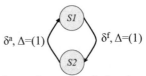

(a) A program that can benefit from node splitting because of dependences cycles.

(b) The dependence graph for the program of (a).

```
            for (i = 0; i < n; i ++){
      S1      a[i] = . . . ;
      S2a     tmp[i] = a[i + 1];
      S2      . . . = tmp[i] + a[i − 1];
            }
```

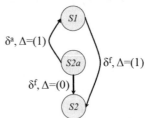

(c) The program of (a) after splitting the nodes and capturing the value of a[i+1].

(d) The dependence graph for the program of (c).

```
            for (i = 0; i < n; i ++) {
      S2a     tmp[i] = a[i + 1];
      S1      a[i] = . . . ;
      S2      . . . = tmp[i] + a[i − 1];
            }
```

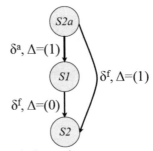

(e) Moving the read of a[i+1] to break the cycle

Figure 4.8: Node splitting example.

the more complicated code of statements $S8$ to $S12$ for the second loop. Given $|T|$ threads, this version will total the partial sums in $\log_2(|T|)$ steps, as shown in Figure 4.9(e) for 16 threads.

To parallelize reductions, a compiler essentially looks for statements of the form $s = s \oplus expr$. Two conditions must be met for the reduction to be parallelized. First, the value of $expr$ must be the same regardless of the loop order it is evaluated in. If $expr$ either only contains variables that are unchanged, or the values are changed in the same iteration of the loop they are used. That is, the flow or anti dependence from write to the read in the reduction statement has direction " or "0" in the loop being parallelized. Second, the left-hand side s must not be used in other statements. Reads of s in some iteration of the parallel loop will get a partial sum that has a different value than it would in a sequential execution of the loop. Figure 4.9(d) gives an example of a loop that violates both of these conditions.

```
S1    for (i = 0; i < n; i ++) {
S2        a[i] = ...;
S3        s = s + a[i];
          }
```

(a) A loop containing a reduction.

```
S1    t = numThreads;
S2    double Sp[t];
S3    parallel for (i = 0; i < n; i ++) {
S4        a[i] = ...;
S5        Sp[t] = s + a[i];
S6    }
S7    for (i = 0;
          i < numThreads; i ++) {
S8        s = s + Sp[t];
S9    }
```

(b) The loop transformed to execute the reduction in parallel.

```
S1    t = numThreads;
S2    double Sp[t];
S3    parfor (i = 0; i < n; i ++) {
S4        Sp[t] = s + a[i];
S5    }
S6    int l2 = log2(numThreads);
S7    step = 1;
S8    for (m = 0; m < l2; m ++) {
S9        for (i = 0;
              i < numThreads;
              i = i + step * 2) {
S10           Sp[i] = Sp[i] + Sp[i + step];
S11       }
S12       step = step * 2;
          }
```

(c) A better parallelization of the loop of (a).

```
S1    t = numThreads;
S2    double Sp[t];
S3    parallel for (i = 0; i < n; i ++) {
S4        s = s + a[i];
S5        a[i − 1] = s;
S6    }
S7    for (i = 0; i < numThreads; i ++) {
S9        s = s + Sp[t];
          }
```

(d) A loop with a reduction operation that cannot be parallelized.

Figure 4.9: Reduction recognition example.

4.9   WHICH TRANSFORMATIONS ARE MOST IMPORTANT?

As mentioned earlier, most loops cannot be directly parallelized because loop carried data dependences exist on the loops. True dependences by their very nature cannot be eliminated—at best the involved statements can be reordered and enforced by barriers, or enforced by producer-consumer synchronization. This is the goal of peeling, splitting and topologically sorting the dependence graph and performing fission. The effectiveness of these transformations is limited by both the form that a computation has been expressed and the data flow properties of the underlying algorithm.

Scalar expansion and array privatization are arguably among the most important transformations a compiler can perform to eliminate dependences. Common programming idioms, such as the use of scalars and arrays within deeply nested loops to hold temporary values, lead to numerous anti and output dependences. Because these dependences are not fundamental to a computation but rather are an artifact of insufficient resources being used, applying transformations that provide additional resources locally within a loop can have a dramatic effect on the number of dependences. It is interesting to observe *register renaming*, which maps architected registers to a combination of architected and unarchitected registers to break output and anti dependences, has eliminated renaming as a transformation in at least some compilers.

CHAPTER 5

Transformation of iterative and recursive constructs

The dependence information gathered using the techniques of Chapter 2 can be used to determine the legality, and sometimes the utility, of a variety of transformations on loops. Many of the transformation we now discuss have as their primary goal improving program performance instead of transforming programs to a form that is easier to parallelize.

5.1 LOOP BLOCKING OR STRIP MINING

Loop blocking is used break the iteration space of a loop into blocks of contiguous iterations. The two primary uses for loop blocking are (i) to split loops up into chunks of computation that can be executed as single vector instruction, and (ii) as one step in performing loop tiling (see Section 5.6) to achieve better locality.

Figure 5.1 shows a loop before and after blocking. The loop is split into a pair of loops, where the outer loop jumps from block to block of the iteration space, and the inner loop traverses the individual iterations in the block. *bs* is the block size, that is, the number of iterations in each block.

```
for (i = 0; i < n; i ++) {
    a[i] = ...;
}
```

```
for (i = 0; i < n; i+ = bs) {
    for (i' = i; i' < n; i' ++) {
        a[i'] = ...;
    }
}
int f = n%bs;
for (i = n − f; i < n; i ++) {
    a[i] = ...;
}
```

(a) A loop to be blocked.   (b) The loop of (a) after blocking.

Figure 5.1: An example of loop blocking.

As can be observed from the code, blocking itself has no affect on the order of memory accesses in the program, and is therefore always legal. The only adverse effect on performance comes from slightly higher loop overheads from the additional inner loop.

5.2   LOOP UNROLLING

Small loop bodies present several problems to a compiler. Compilers are limited in their ability to express optimizations that increase cross-iteration sharing of data values[1]. Other problems with small loop bodies include a smaller window over which to do instruction scheduling and there are fewer instructions over which loads can be moved to mask memory latency and a smaller window over which register allocation can be performed, potentially leading to additional loads and stores of data value.

Figure 5.2 demonstrates the benefits of loop unrolling. The computation shown is a simplified stencil. Each iteration of the original loop performs loads of a[i], a[i − 1] and a[i + 1], and stores into a[i]. The value of a[i] stored in an iteration is the value of a[i − 1] in the next iteration, and the value of a[i + 1] in an iteration is the value of a[i] in the next iteration. Therefore, it would be more efficient to reuse the values already in registers in the next iteration, and only have to load a new value of a[i + 1] instead of having to load values of a[i], a[i − 1] and a[i + 1] in each iteration. Code to do this is shown in Figure 5.2(c). As can be seen from the code in Figure 5.2(c) many common array accesses exist across statements in the unrolled loop, which would have been in separate iterations in the original loop. Register allocating across the body of the unrolled loop makes it easy to exploit these common data accesses, whereas in the rolled loop tracking register uses across iterations would have been very difficult.

Unrolling also opens up a wider window for other optimizations to occur. Thus, in the loop shown, the sum a[i] + a[i + 1] is the sum a[i − 1] + a[i] in the next iteration, and the temporary holding the first value can be reused for the second value. The number of compares and increments of the loop index variable are also reduced, making the loop overhead smaller. At run time, the larger number of operations between branches can also increase the window over which out-of-order processors can schedule code without speculating.

Because loop unrolling does not change the order of fetches and stores, it is always legal to perform.

5.3   LOOP FUSION AND FISSION

Loop fusion and fission (also called *loop distribution* in the literature [250]) are inverse transformations that merge multiple loops into a single loop, and split a single loop into multiple loops. An example of this is shown in Figure 5.3.

Loop fusion produces a program that is faster for several reasons: (i) with lower loop overheads only one loop starts, and since the number of iterations executed is one-half of the total for the two unfused loops, the number of index variable increments and checks is also halved; and (ii) temporal locality is better since accesses to common storage in the two loops will be done temporally closer together.

---

[1] Software pipelining, discussed in Section 3.7, gives one way of exploiting cross-iteration reuse of values.

```
for (i = 0; i < n; i ++) {
S1     a[i] = a[i − 1] + a[i] + a[i + 1];
}
```

(a) A loop that can benefit from unrolling.

```
for (i = 0; i < n; i+ = 4) {
S1      a[i] = a[i − 1] + a[i] + a[i + 1];
S1a     a[i + 1] = a[i] + a[i + 1] + a[i + 2];
S1b     a[i + 2] = a[i + 1] + a[i + 2] + a[i + 3];
S1c     a[i + 3] = a[i + 2] + a[i + 3] + a[i + 4];
}
int f = n%4;
for (i = n − f; i < n; i ++) {
S1      a[i] = a[i − 1] + a[i] + a[i + 1];
}
```

(b) The loop of (a) after unrolling.

Figure 5.2: An example of loop unrolling.

Loop fission can be used to create loops where one operation exists per loop, enabling vectorization. As seen in the loop fission example of Figure 5.3(a), when the loop is distributed the dependence between $S1$ and $S2$ is enforced by the sequential execution of the two loops. If both of the resulting loops, shown in Figure 5.3(c), are parallelized or vectorized, the barrier after each loop will enforce the dependence.

Because fusion and fission change the order in which loads and stores are performed, it is necessary to check the dependence relation before and after performing the transformation to determine if it is legal. In particular, when performing loop fission, every flow dependence from some statement S_p to S_q should be replace by a def-use edge from S_p to S_q, every anti-dependence from S_p to S_q should be replaced by a use-def edge from S_p to S_q, and every output dependence from S_p to S_q should be replace by a def-def edge from S_p to S_q. Thus, in the example of the figure, the flow dependence from $S1$ to $S2$ is replaced by a def-use chain from $S1$ to $S2$ in the new loop. We note that a lexically backwards flow (anti) dependence (i.e., consider if statements $S1$ and $S2$ were re-ordered in Figure 5.3(a)) will become a lexically forward def-use (use-def) edge after fission, and a lexically backward output dependence will become a lexically forward def-def edge with the source and sink reversed, thus violating the legality constraints. As described in Chapter 3.2, acyclic portions of the dependence graph may be sorted so that dependences are lexically forward, with a legal fissioning then being possible.

A legal fusion requires that all def-use (use-def/def-def) edges from S_p to S_q in the loops being fused should become with flow (anti/output) dependences from S_p to S_q, and no other dependences (other than those that existed in the unfused loops) should exist in the fused loop.

```
for (i = 0; i < n; i++) {
    a[i] = ...;
    ... = a[i − 1];
}
```

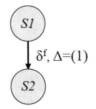

$\delta^f, \Delta=(1)$

(a) The loop of (c) after fusion.

(b) A dependence graph for the loop.

```
for (i = 0; i < n; i++) {
    a[i] = ...;
}
for (i = 0; i < n; i++) {
    ... = a[i − 1];
}
```

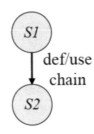

def/use
chain

(c) The loop of (a) after fission or loop distribution.

(d) A dependence graph for the loop of (c).

Figure 5.3: Examples of loop fusion and fission.

5.4   LOOP REVERSAL

Loop reversal is a transformation that reverses the order in which the iterations of the loop are executed. Thus, a loop that originally runs from $0 \ldots n - 1$ will now run from $n - 1 \ldots 0$, as shown in Figure 5.4. Figures 5.4(a) and (b) show the original pair of loops to be fused, and the use-def information for the loops. Figures 5.4(c) and (d) show the fused loop and the dependence graph for the loop. Observe that the Use/Def chain from $S1$ to $S2$ in the loop of Figure 5.4 has become a flow dependence in the fused loop, meaning the order of reads and writes of a elements has changed after fusing, and the fusion is illegal.

When the iteration space of a loop is reversed, the direction of dependences within that reversed iteration space are also reversed. Thus, a "<" dependence becomes a ">" dependence, and vice versa. More intuitively, in the notation expressing the dependence as the sign of the direction, the direction of dependences in the reversed loop are found by multiplying the direction in the original loop by -1. This effect of loop reversal can be used to fuse loops that might otherwise be illegal to fuse. After reversing the iteration space of the loop shown in Figure 5.4(c), the loop of Figure 5.4(e) is produced. Reversing the direction gives a flow dependence from $S2$ to $S1$ with a direction of "-1" or ">". Now, since dependences must go from earlier in the loop execution to later, these directions cannot be correct; what has happened is that the flow dependence from $S2$ to $S1$ has become an anti dependence from *S1* to *S2* with a direction of "1" or "<". This direction corresponds to the flow of data implied by the Use/Def information shown in Figure 5.4(b), and therefore the fusion is now legal.

```
for (i = 0; i < n; i++) {
    ... = a[i – 1];
}
for (i = 0; i < n; i++) {
    a[i] = ...;
}
```

(a) A pair of loops to be fused.

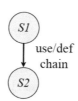

(b) The Use/Def chains for the loops of (a).

```
for (i = 0; i < n; i++) {
    ... = a[i – 1];
    a[i] = ...;
}
```

(c) The loop of (a) after fusing.

(d) The dependence graph for the fused loop of (c).

```
for (i = n – 1; i >= 0; i – –) {
    ... = a[i – 1];
    a[i] = ...;
}
```

(e) The loop of (a) after fusing and reversing the loop order.

(f) The dependence graph for the loop of (e).

Figure 5.4: An example of how loop reversal aids in fusion.

5.5 LOOP INTERCHANGE

Loop interchange is a transformation that increases the temporal and spatial locality of array accesses. The transformation takes two loops - i_{out} and i_{in}, where i_{out} (i_{in}) is an outer (inner) loop, and interchanges the positions of i_{out} and i_{in} so that i_{in} is the outer loop.

There are two benefits from applying loop interchange. First, the locality, and therefore the cache behavior, of array accesses can be improved by interchange. Languages define the order that arrays are laid out in memory. Fortran column elements that are adjacent in the Cartesian space defined by the array are stored adjacent to one another in memory, while C, C++ and Java make row elements adjacent in memory[2]. Array accesses may occur in the wrong order because of library code being reused for another language (for example, Fortran arrays have adjacent column elements next to each other in memory, whereas C has adjacent row elements adjacent to each other in memory), the literal translation of algorithms from one language to another, and from transcribing algorithms in a book that might be written to emphasize intuitiveness without regard for locality effects. By interchanging the loop involved in indexing the arrays in the wrong order, the access order is changed to one with better cache behavior. If the loop has array accesses that occur in the proper order in dimensions indexed by the loops to be interchanged, they will now perform their accesses in the wrong order. In these cases, *tiling*, discussed in Section 5.6, should be used.

Second, if an inner loop is parallel, interchanging it with an outer loop will decrease the number of times the parallel loop is instantiated at run time and will increase the amount of work in the parallel loop. Both of these reduce the overhead of exploiting the parallelism inherent in the loop.

Because loop interchange affects the order of accesses—accesses that previously traversed elements in row order will now traverse them in column order, and vice versa—dependence information must be used to insure that the transformation is legal to apply. Figures 5.5(a)–(c) show a double nested loop that is a candidate for interchange, an iteration space diagram for the loop, and the dependence graph for the loop. As can be seen, there is a flow dependence with a "$[1, -1]$" ("$[<, >]$") direction vector on the loop. When interchange is applied to the loop nest, the direction vector elements associated with the interchanged loops are also interchanged. Figures 5.5(c) and (d) show the interchanged loop, ISD and dependence graph. After interchange, the flow dependence would have a direction vector of "$[-1, 1]$" ("$[>, <]$"). This, however, would imply that the dependence would travel backwards in the iteration space, which cannot happen since dependences always travel forward in the loop nest's iteration space. Therefore, the interchanged loop now has an *anti* dependence running from the read to the write of a with a direction of "$[1, -1]$" ("$[<, >]$"). In simpler terms, this means that the read of an element of a now occurs after, rather than before, the write. Because these accesses occur in a different order than in the original program, the transformed program may give a different (and wrong) result.

Figures 5.6(a)–(d) show another loop nest and its iteration space diagram before, and after, performing loop interchange. In this case, there is a flow dependence with a direction of "$[1, 0]$" ("$[<, =]$"). After interchange, the direction vector becomes "$[0, 1]$" ("$[=, <]$"), which indicates a flow dependence that is still forward in the iteration space of the loop. Thus, the interchange is legal.

For a two-deep loop nest, the possible direction vectors and what they become after interchange are:

---

[2]Java does not specify a layout for arrays, and prohibits programs from depending on the layout of arrays, but in practice adjacent array elements are contiguous in memory. As well, Java does not define two dimensional arrays but builds up multi-dimensional arrays from arrays of *references* or pointers, as shown in Figure 2.9. This effectively makes adjacent row elements contiguous.

$$
\begin{bmatrix}
\text{``}[0,0]\text{''}\ (\text{``}[=,=]\text{''}) & \Rightarrow & \text{``}[0,0]\text{''}\ (\text{``}[=,=]\text{''}) \\
\text{``}[0,1]\text{''}\ (\text{``}[=,<]\text{''}) & \Rightarrow & \text{``}[1,0]\text{''}\ (\text{``}[<,=]\text{''}) \\
\text{``}[1,-1]\text{''}\ (\text{``}[<,>]\text{''}) & \Rightarrow & \text{``}[-1,1]\text{''}\ (\text{``}[>,<]\text{''}) \\
\text{``}[1,0]\text{''}\ (\text{``}[<,=]\text{''}) & \Rightarrow & \text{``}[0,1]\text{''}\ (\text{``}[=,>]\text{''}) \\
\text{``}[1,1]\text{''}\ (\text{``}[<,<]\text{''}) & \Rightarrow & \text{``}[1,-1]\text{''}\ (\text{``}[<,>]\text{''})
\end{bmatrix}
$$

The direction vectors "$[-1,1]$" ("$[>,<]$"), "$[-1,0]$" ("$[>,=]$") and "$[-1,-1]$" ("$[>,>]$") are not possible because they imply the dependence sink executes before the source. All of the possible direction vectors except for "$[1,-1]$" ("$[<,>]$") yield legal vectors when interchanged, and therefore can exist on the pair of loops to be interchanged. The "$[1,-1]$" ("$[<,>]$") is the only direction vector that precludes loop interchange.

The discussion above has assumed that the interchange involves the outermost loop. If it does not, and the direction vectors of all dependences after the interchange are lexicographically greater than $[0,0,\ldots,0]$, the interchange is legal. In practice, this means that if for every direction vector there is at least one "$<$" element corresponding to a loop that is outside of the interchanged loops in the loop nest, the interchange is legal. Thus, in a triply nested loop, if the only dependence has a direction vector of "$[1,1,-1]$" ("$[<,>]$"), after interchange the direction vector would be "$[1,-1,1]$" ("$[<,>,<]$"), and the interchange is legal.

5.6   TILING

When a loop contains accesses to arrays that are being performed in an inefficient order, and accesses that are being performed in the correct order, interchange will enhance the locality for the accesses happening in the wrong order, and will degrade the locality of the accesses already happening in the efficient order. This situation occurs, for example, in matrix multiply and array transpose.

Consider the loop nest shown in Figure 5.7. The $a[i_1, i_2]$ reference accesses the array in row-major order, and is locality friendly, as shown by the horizontal, dark arrows in Figure 5.7(c). The $a[i_2, i_1]$ reference accesses the array in column-major order, as shown by the vertical, lighter arrows. Clearly, interchange alone will not improve the overall locality of the accesses. What is desired is to *tile* the iteration space with non-overlapping sub-spaces that both entirely cover the original iteration space, and reorder the accesses such that all of the array regions accessed in a tile fit into the cache.

Figure 5.7(c) shows a symbolic sequence of tiles being accessed by the program of Figure 5.7(a) after tiling (the top row) and the sequences of the a array accessed by each of tiles (the bottom row). After tiling, accesses to the array region for a single reference are occur in a contiguous set of iterations of the loop nest, i.e., they are temporally co-located.

The first step in producing code that gives this behavior is to block each loop (see Section 5.1), as shown in Figure 5.7(d). Blocking does not affect the order of accesses and is always legal. To make the accesses occur in a more cache friendly manner, a second step is performed, i.e., the i_1' and i_2 loops are interchanged. This leaves the outer loops in the loop nest (or tiled portion of the loop

```
for (i₁ = 0; i₁ < n; i₁ ++){
    for (i₂ = 0; i₂ < n; i₂ ++) {
        a[i₁, i₂] = ...
        ... = a[i₁ − 1, i₂ + 1]
    }
}
```

(a) A loop nest that is a candidate for loop interchange.

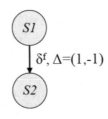

(b) The iteration space diagram for the loop nest of (a).

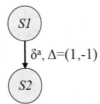

(c) The dependence graph for the loop nest of (a).

```
for (i₂ = 0; i₂ < n; i₂ ++) {
    for (i₁ = 0; i₁ < n; i₁ ++) {
        a[i₁, i₂] = ...
        ... = a[i₁ − 1, i₂ + 1]
    }
}
```

(d) The loop nest of (a) with the loops interchanged.

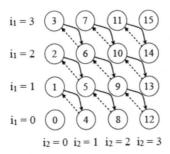

(e) The iteration space diagram for the interchanged loop nest of (d). Dashed lines show the flow of data in the original loop order and solid lines the flow of data after loop interchange.

(f) The dependence graph for the interchanged loop nest of (d).

Figure 5.5: An example of where loop interchange is illegal.

```
for (i₁ = 0; i₁ < n; i₁ ++) {
    for (i₂ = 0; i₂ < n; i₂ ++) {
        a[i₁, i₂] = ...
        ... = a[i₁ − 1, i₂ − 1]
    }
}
```

(a) A par of loops for which interchange is legal.

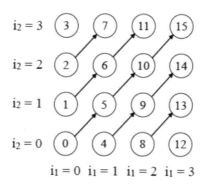

(b) The iteration space diagram for the loop of (a).

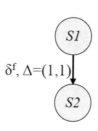

(c) The dependence graph for both the original loop nest of (a), and the loop nest after interchanging the i₁ and i₂ loops.

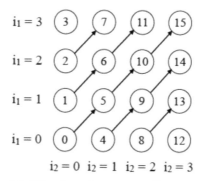

(d) The iteration space diagram for the loop nest of (a) after interchange.

Figure 5.6: An example of where loop interchange is legal.

nest, if there are other loops surrounding the tiled loops) being the i_1 and i_2 loops that iterate over the different blocks, or tiles, in the nest. The inner loops are the i_1' and i_2' loops that perform the iterations within each tile. Thus, the iteration space visits each tile in turn, and then traverses the iterations that perform accesses within the tile. This in turn causes each region of the array accessed by each tile to be visited, leading to a cache friendly behavior. We note that tiling is legal whenever the interchange used in the second step of the transformation is legal, as described in Section 5.5.

The tiled code that achieves this behavior is shown in Figure 5.7(c).

```
for (i₁ = 0; i₁ < n; i₁ ++) {
  for (i₂ = 0; i₂ < n; i₂ ++) {
    a[i₁, i₂] = a[i₂, i₂];
  }
}
```

(a) A program with locality friendly and hostile accesses.

```
for (i₁ = 0; i₁ < n; i₁+ = bsᵢ) {
  for (i′₁ = i₁; i′₁ < i₁ + bsᵢ₁; i′₁ ++) {
    for (i₂ = 0; i₂ < n; i₂+ = bsᵢ₂) {
      for (i′₂ = i₂; i′₂ < i′₂ + bsᵢ₂; i′₂ ++) {
        a[i₁, i₂] = a[i₂, i₁]
      }
    }
  }
}
```

(b) After loop blocking – the first phase of tiling.

```
for (i₁ = 0; i₁ < n; i₁+ = bsᵢ₁) {
  for (i₂ = 0; i₂ < n; i₂+ = bsᵢ₂) {
    for (i′₁ = i₁; i′₁ < i₁ + bs; i′₁ ++) {
      for (i′₂ = i₂; i′₂ < i₂ + bsᵢ₂; i′₂ ++) {
        a[i′, i′₂] = a[i′₂, i′];
      }
    }
  }
}
```

(c) After strip mining – the second phase of tiling.

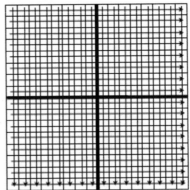

(d) A graphical representation of the access patterns before tiling. Each square represents an element of the array, and the order of row and column accesses of the two references to a are shown as arrows.

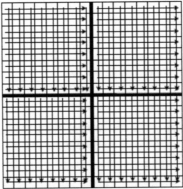

(e) A graphical representation of the access patterns after tiling. Each square represents an element of the array, and the order of row and column accesses of the two references to a are shown as arrows.

Figure 5.7: An example of tiling and its effect on the execution order and array access patterns.

5.7 UNIMODULAR TRANSFORMATIONS

Some, but not all, loop transformations are linear transforms of the iteration spaces and loop bounds. In particular, skewing, reversal and interchange can be represented as such, while tiling, strip mining, fusion, and distribution cannot. However, as we have seen, skewing can change the distances of dependences within a loop nest, and enable these other optimizations. These observations led to the development of a theory of *Unimodular loop transformations* [21, 242, 243].

```
for (i₁ = 0; i₁ < n; i₁ ++) {
    for (i₂ = 0; i₂ < n; i₂ ++) {
        a[i₁, i₂] = a[i₁, i₂] + a[i₁ − 1, i₂ + 2];
    }
}
```

Figure 5.8: A loop whose performance can be improved by Unimodular transformations.

Consider the loop nest of Figure 5.8 with an outer i_1 and inner i_2 loop. The loop nest can be represented by the vector:

$$\begin{bmatrix} i_1 \\ i_2 \end{bmatrix}$$

and the interchange of the two indices can be represented by the matrix vector operation

$$\begin{bmatrix} 0 & 1 \\ 1 & 0 \end{bmatrix} \begin{bmatrix} i_1 \\ i_2 \end{bmatrix} = \begin{bmatrix} i_2 \\ i_1 \end{bmatrix},$$

i.e., by multiplying the vector of loops by the appropriate permutation matrix. To model the effect of the interchange on the distance vectors of the dependence, the distance vectors can be multiplied by the same permutation matrix. In the example of Figure 5.8, the dependence distance vectors are $[0, 0]$ and $[1, −2]$. Multiplying the transposed dependence vectors by the permutation matrix yields

$$\begin{bmatrix} 0 & 1 \\ 1 & 0 \end{bmatrix} \begin{bmatrix} 0 & 1 \\ 0 & −2 \end{bmatrix} = \begin{bmatrix} 0 & −2 \\ 0 & 1 \end{bmatrix}$$

or distance vectors of $[0, 0]$ and $[−2, 1]$. The $[0, 0]$ vector is legal, but the $[−2, 1]$ is lexicographically less than $[0, 0]$ and indicates a dependence traveling backwards in the iteration space of the loop. Thus, the interchange is illegal.

In a similar fashion, loop reversal (see Section 5.4) can be expressed as a matrix operation. Again, using the loop of Figure 5.8 as an example, and reversing the inner i_2 loop implies the system

$$\begin{bmatrix} 1 & 0 \\ 0 & −1 \end{bmatrix} \begin{bmatrix} 0 & 1 \\ 0 & −2 \end{bmatrix} = \begin{bmatrix} 0 & 1 \\ 0 & 2 \end{bmatrix}$$

or distance vectors of $[0, 0]$ and $[1, 2]$. Neither distance $[0, 0]$ vector is lexicographically less than less than $[0, 0]$, and therefore the reversal is legal.

Finally, skewing can be represented by a identity matrix with elements representing the skewing factor. Thus, the matrix

$$\begin{bmatrix} 1 & 0 \\ 2 & 1 \end{bmatrix}$$

represents skewing the iteration space by twice the distance on the outer i_1 loop, and in the example loop would yield distance vectors of $[0, 0]$ and $[2, 0]$.

The above examples assume the dependence distances are known and constant. It is sometimes the case that a dependence direction is known, but the distance is not. The above framework can handle this in several steps. First, "compound" distances such as $[*, *]$ are represented as a pair of directions $[0, +]$ and $[+, \pm]$. This prevents the impossible dependence direction $[-, \pm]$ from being considered in the following. Next, the directions can be represented as a ranges of integers, with the two directions above being represented as $[0 : 0, 1 : \infty]$ and $[1 : \infty, -\infty : \infty]$. An arithmetic can be defined on these vectors of rangers. Let $l_1, l_2, u_1, u_2 \in \mathbb{Z}$, then $l_1 : u_1 + l_2 : u_2 = l_1 + l_2 : u_1 + u_2$, and $s \cdot l : u = s \cdot l : s \cdot u$ when $s \geq 0$ and $s \cdot l : u = s \cdot u : s \cdot l$ otherwise. Multiplication or addition of any k and $\infty(-\infty)$ yields $\infty(-\infty)$. Using this representation, directions can be converted to ranges and manipulated similar to scalars. Loops that have ∞ cannot be parallelized, since no skewing factor exists that eliminates the dependence and allows overlap of execution.

The algorithms presented in [21, 242, 243] show how to efficiently find a Unimodular matrix T that represents the desired and combined effects of skewing, reversal and interchange to give the desired loop nest. Manipulation of the T matrix allows computation of the loop bounds for the new nest. While the use of Unimodular transformations does not give greater power than a framework directly applying skewing, reversal and interchange, it does allow these to be done by manipulating matrices rather than the compiler IR, and therefore allows simpler and easier to maintain approaches within the compiler to represent and perform these transformations.

CHAPTER 6

Compiling for distributed memory machines

Up until now, our focus has been almost entirely on compiling for parallelism on shared memory machines, motivated in part by the dominance of multicore processors. Nevertheless, all large machines for high-performance numerical computing, as well as many small scale clusters, have a physically distributed memory architecture and are programmed using a distributed memory programming model.

Distributed memory machines consists of *nodes* connected to one another by using Ethernet or a variety of proprietary interfaces. Examples of these include the IBM Blue Gene machines [29, 162], clusters of workstations and small clusters of rack mounted nodes [173]. Distributed memory machines usually execute *single program multiple data* (SPMD) programs. In an SPMD program, the program's data is distributed (or spread) across multiple instances of the same program, allowing each instance of the program to compute in parallel on its part of the data. Because the computations are not fully independent, communication must sometimes be used to transmit the coherent copy of a datum to a process needing that copy. This communication is done across the nodes of a distributed memory machine using explicit *messages* and *collective communication* that must be specified by either the programmer or the compiler.

It is increasingly common for nodes to be multicore processors. In this case, each core is usually treated as a single node, and the fact that a shared memory exists across some of the nodes is exploited by the underlying MPI library, if at all. In some cases a hybrid model is used, where a multithreaded program runs on each nodes and communicates via message passing with other nodes, but we do not know of compiler support for this model. The complexity of maintaining two programming models has prevented this from becoming standard practice. In what follows, we assume that each core is treated as a different node.

Figure 6.1(b) shows a high-level diagram of a distributed memory machine, and Figure 6.1(a) shows a simple program that computes the average of the elements in an array. Several interesting concepts are touched on by this figure. For example, the network topology is not specified. In general, compilers view data as close and far, and are not cognizant of the details of specific network topology. The figure also shows that arrays are distributed across the processors while a copy of each scalar often resides on every process.

In Figure 6.1(c), the SPMD program that performs this computation is shown. For simplicity, it is assumed that the data array has 100 elements, and that there are four processes, i.e., the array

```
float a[4 * n], s, avg;

...
avg = 0; s = 0;
for (i = 0; i < n; i++) {
    s = s + a[i];
}
avg = s/(float)n;
```

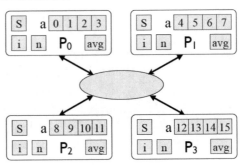

(a) A sequential program that computes the average of 4*n numbers.

(b) The data of (a) distributed across four processes. For clarity, the indices of a are shown for the original undistributed array.

```
#include "mpi.h"
#include < stdio.h >
#define SIZE 4

int main(argc, argv)
int argc;
char * argv[];
int numtasks, rank, count;
float a[n];
int n;

MPI_Init(&argc, &argv);
MPI_Comm_rank(MPI_COMM_WORLD,
        &rank);
```

```
MPI_Comm_size(MPI_COMM_WORLD,
        &numtasks);

avg = 0; s = 0;
for (i = 0; i < n; i++) {
    s = s + a[i];
}

MPI_Allreduce(&s, &s, 1, datatype,
        op, comm)
avg = s/(float)4 * n;

MPI_Finalize();
```

(c) A distributed memory SPMD program that uses MPI to perform the computation of the program of (a) and the data distribution of (b).

Figure 6.1: An example of a distributed memory program.

size is evenly divisible by the number of processes. In forming the SPMD program, several problems must be solved:

Data distribution. The data used by the program must be distributed across the different processes and placed into *local arrays*, and the subscripts accessing the original, undistributed arrays must be modified to access the local, distributed arrays. These issues are discussed in detail in Section 6.1.

Computing the data accessed by a reference.   As a precursor to efficiently solving the next two problems, it is necessary to compute the data accessed by an array reference. How to do this is discussed in Section 6.2.

Computation partitioning.   The bounds of the loops on each processor must be shrunk to only cover the iterations of the computation that is performed by each process. This problem is discussed in Section 6.3.

Generating communication.   A global, coherent view of the data is maintained by explicit communication between the program instances running on different processes. As well, it is desirable to move communication out of inner-most loops, both to maximize parallelism and to minimize the overhead of communication operations. How compilers do this is discussed in Section 6.4.

Programming languages targeting distributed memory machines.   Several programming languages have been designed to support program development for distributed memory machines. Three of the more popular are discussed in Section 6.5.

6.1   DATA DISTRIBUTION

Parallelism on distributed memory machines is exploited by spreading the computation across multiple processors or cores. On distributed memory machines, this is done by distributing the data across the processors executing the program. This has two benefits. First, each processor (or more accurately, each process executing on a processor) computes the new values for the data that resides on it, or is *owned* by it, in parallel with the other processes. The scheduling of computation on the process that owns the storage location whose value is being computed is called the *owner computes rule* [198]. Second, by distributing the large arrays, programs can scale both in performance and in their maximum size, since each new process added to the computation increases the amount of physical memory the program can use.

There are five commonly used data distribution mechanisms.

1. *Replicated:* a copy of the data is given to every process. Figures 6.2(a) and (b) shows an example of a replicated distribution of an array.

2. *Block:* The array is divided into $|P|$ chunks (where $|P|$ represents the number of processes), or blocks, of contiguous elements which are distributed onto the processes. Figures 6.3(a) and (b) show an example of a block distributed array.

3. *Cyclic or cyclic(1):* Elements of the array are distributed in a round-robin, or cyclic, fashion across processes. Figures 6.4(a) and (b) show an example of a cyclically distributed array.

4. *Cyclic(bs):* The array is broken into blocks containing *bs* (block size) elements, which are then distributed in a cyclic pattern across the processes. This distribution is sometimes referred

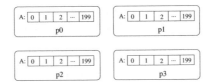

Figure 6.2: Layout of the replicated C array a[99] on three processes.

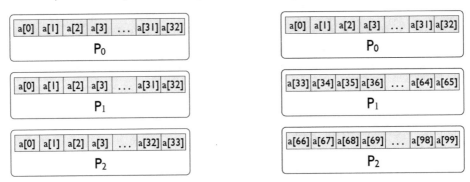

(a) The local storage for a block distribution of the C array a[99] distributed onto three processes.

(a) The elements of a[99] in a block distribution onto three processes. Elements are numbered using their global indices.

Figure 6.3: Layout of a block distributed array onto three processes.

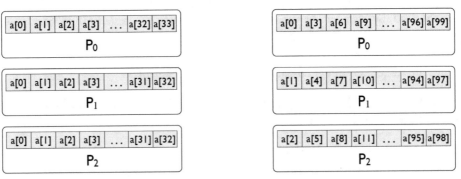

(a) The local storage for a cyclic distribution of the C array a[99].

(b) The elements of a[99] of (a) in a cyclic distribution on three processors. Elements are numbered using their global indices.

Figure 6.4: Layout of a cyclically distributed array onto three processes.

to as a *block-cyclic* distribution. Figure 6.5(a) and (b) show an example of a block-cyclically distributed array.

5. *General:* Generalizations of block-cyclic distributions exist where the array is broken into varying sized blocks, but are beyond the scope of this lecture and are not supported by any commercially available or widely used compilers. In Chapter 8, papers that explore this in more depth are suggested.

Each dimension of an array is distributed independently of other dimensions. Thus, one dimension can be distributed with a block distribution, and another with a cyclic, block-cyclic or replicated distribution. The distributed array will have two sets of bounds: the *global bounds*, which are the bounds of the initial undistributed array, and the *local bounds*, which are the bounds of the array on some process. We now describe each distribution in detail for a one-dimensional array a with global lower and upper bounds of L_a and U_a, respectively.

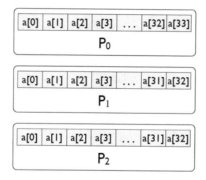

(a) The local storage for a block-cyclic distribution of the C array a[99].

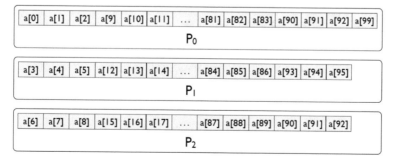

(b) The elements of a[99] of (a) in a block-cyclic distribution on three. processes. Elements are numbered using their global indices.

Figure 6.5: Layout of a block-cyclic distributed array onto three processes with a block size of three.

6.1.1 REPLICATED DISTRIBUTION

The replicated distribution is the simplest of the five distributions we will discuss. Each process p owns elements $\mathcal{O} = \{e | L_a \leq e < U_a\}$. Element e is owned by all processes $\{p_i | 0 \leq i < |P|\}$. Because all processes own every element of the array, under the owner computes rule all processes will (redundantly) compute new values of all elements. and will use these locally computed values. This means that if some element of a replicated array a, e.g., a[i], is assigned a value that is a function of the process id, e.g., a[i] = pid, then the values of a[i] that exist, and are read on different processes will be different. Replicating an array is desirable when the cost of performing the redundant computation is less than the cost of communicating the subset of values computed on each process to all other processes.

6.1.2 BLOCK DISTRIBUTION

In the block distribution, the array needs to be divided into $|P|$ blocks. We present a computationally simple way of describing this from [187]. Let L_a be the lower bound of a, and let U_a be the upper bound of a for the global bounds. The number of elements in a is then $|a| = U_a - L_a + 1$. The formula for $L_{a,p}$ ($U_{a,p}$), the lower (upper) bound of a on process p is:

$$L_{a,p} = \left\lfloor \frac{p\,|a|}{|P|} \right\rfloor$$
$$U_{a,p} = \left\lfloor \frac{(p+1)\,|a|}{|P|} \right\rfloor - 1 .$$

Note that the upper bound is simply one less than the lower bound of the next highest processor.

The elements owned by are process p can be described by the triplet $[L_{a,p} : U_{a,p} : 1]$. The set $\mathcal{O} = \{e | L_{a,p} \leq e \leq U_{a,p}\}$. The owning process of some element $a[\sigma(I)]$, where $\sigma(I)$ is the subscript for the block distributed dimension whose bounds are being computed and I is the index variables for the loop nest surrounding the reference, is

$$owner(e) = \left\lfloor \frac{|P|\,(\sigma(I)+1) - 1}{n} \right\rfloor .$$

Because computing the intersection of different owned sets requires knowledge of other distributions, we defer this discussion until Section 6.4.

6.1.3 CYCLIC DISTRIBUTION

The cyclic distribution spreads elements of a across processes in a round robin fashion. Process p has the p'th element of the array, and there are $U_a - L_a + 1$ array elements total. Thus, after placing the first element onto process p, there are $U_a - L_a + 1 - (p + 1)$ elements left to distributed, and

p will get every $|P|$'th of those elements. More succinctly, the number of elements $|e|$ is given by:

$$|e| = \left\lfloor (U_a - L_a - p)/|P| \right\rfloor + 1. \tag{6.1}$$

Thus, the maximum element index on the process, in the global index space of the array, is

$$p + |P|(|e| - 1) . \tag{6.2}$$

The bound on the number of elements on each process allow us to define the set of elements owned by each process as a Diophantine expression with bounds, i.e.,

$$\mathcal{O} = \{p + j \cdot |p| \mid 0 \le p < |e|\}.$$

6.1.4 BLOCK-CYCLIC DISTRIBUTION

The block and cyclic distributions are degenerate forms of the more general block-cyclic distribution. Whereas the block distribution emphasizes spatial locality, and the cyclic distribution emphasizes load balance, the block-cyclic distribution allows both benefits to be achieved. In the block-cyclic distribution, blocks of size bs are cyclically distributed across the $|P|$ processes.

Historically, the block-cyclic distribution was the last distribution for which access and ownership functions were defined in a closed form. The problem was of great concern in the early nineties, because High Performance Fortran [124], which sought to automate much of the difficulty of writing distributed memory programs, specified the block-cyclic distribution as one of its fundamental and supported distributions. The difficulty of creating closed formed expressions for the block-cyclic distribution were, in hindsight, overstated. The crux of the problem is that while the elements owned by a process for an array that is block or cyclically distributed can be expressed as a linear function in one variable, the block-cyclic distribution requires the use of two variables. Once this key fact is observed, a solution is straightforward to obtain.

Because of this, there were no good solutions for block-cyclic distributions until the mid-1990s, at which time multiple solutions appeared almost simultaneously. A good overview of these solutions can be found in [239]. Most of these solutions varied in performance by a small constant, and in most cases the performance differences can be attributed to implementation tuning. In the remainder of this section we will focus on one of these techniques, that of [153, 154], because of the author's familiarity with the method.

Figures 6.5(a) and (b) show the blocks of data distributed. The technique described below will use one variable to traverse the blocks that are owned by a process, and a second variable to traverse the array elements within a block.

We will label the blocks resident on the process by the first element, in the global index space of the array, of each block. Then, the blocks β_p owned by process p are given by

$$\beta_p = [bs \, p : U_a - 1 : bs|P|] ,$$

where each block is labeled by the index of the first element in the block with the local array bounds. To enumerate the elements within a block $b \in \beta_p$ we use

$$E = [0 : bs - 1 : 1]$$

and the region (or elements) of an array owned by a process p is

$$\mathcal{O}_{a,p} = \{b + e \mid \forall b \in \beta_p, e \in E\}.$$

If we label each block on a process p from $L_{b,p} = 0$ to some upper bound $U_{b,p} - 1$, that is the number of blocks, then

$$U_{b,p} = \left\lceil \frac{U_a - bs\, p}{bs|P|} \right\rceil.$$

We can rewrite $\mathcal{O}_{a,p}$ as

$$\mathcal{O}_{a,p} = \{bs|P|b + bs\, p + e | L_{\beta,p} \leq b \leq U_{\beta,p}, 0 \leq e \leq bs - 1\} . \tag{6.3}$$

6.2   COMPUTING THE REFERENCE SET

The next problem is how to compute the set of elements owned by some process p—the set $\mathcal{O}_{a,p}$ of elements of some array a—that are accessed by a $a[\ldots, \sigma_d(I), \ldots]$, where $\sigma_d(I)$ is the subscript of the block-distributed dimension d of a under consideration. I is the vector of normalized (see Section 4.2) index variable values in the loop nest surrounding the reference, and each $i_j \in I$ takes on values $0 \leq i_j < U_{i_j}$. It is desirable to be able to compute the elements of dimension d that are both accessed by this reference and that reside on the process.

Let $\sigma_d(I) = c_1 i_1 + c_2 i_2 + \ldots + c_n i_n + c_o$, then the region of the array referenced by $\sigma_d(I)$ across every process is

$$\rho_d = \{\Sigma^n j = 1(c_j i_j) + c_0 | \forall i_j, 0 \leq i_j \leq U_{i_j}\} . \tag{6.4}$$

We are now ready to compute the array elements accessed in the dimension d of array reference $a[\sigma_d(I)]$ on process p, using some distribution. Intuitively, we need to intersect the set $\mathcal{O}_{a,p}$ of elements on process p under the distribution with the set ρ_d of elements accessed by the reference. This can be done by intersecting the affine expression for $\mathcal{O}_{a,p}$ with the array elements accessed by the affine subscript expression σ, i.e., by solving the Equations 6.3 and 6.4.

Because block-cyclic is the most complicated distribution, we will examine computing the reference set for it in detail. For simplicity, we assume that $\sigma_d(I)$ only involves a single loop index i_j.

For a block cyclic distribution, we again equate the expressions for what is accessed by the reference and what is owned by the process, getting:

$$c_j i_j + c_0 = bs|P|b + bs\, p + e.$$

Moving terms not involving the equation variables e, i_j and b to the left-hand side gives:

$$c_0 bs - bs\ p = bs|P|b + e - c_j i_j.$$

We can solve this equation for the intersection using the techniques for solving Diophantine equations discussed in Section 2.3.2 and in more detail in Chapter 7. We first form the C matrix:

$$C \quad = \quad [c_0 bs - bs\ p]$$

and the A matrix:

$$A = \begin{bmatrix} bs|P| \\ 1 \\ -c_j \end{bmatrix}$$

Forming the 2×4 IA matrix and using the elimination procedure noted in Section 7.1 gives:

$$\begin{bmatrix} 1 & 0 & 0 & bs|P| \\ 0 & 1 & 0 & 1 \\ 0 & 0 & 1 & -c_j \end{bmatrix} \Rightarrow \begin{bmatrix} 0 & 1 & 0 & 1 \\ 0 & c_j & 1 & 0 \\ 1 & -bs|P| & 0 & 0 \end{bmatrix}$$

where the first three columns are the U matrix and the last column is the D matrix. Since $TD = C$ we can write T as $T = [c_0 bs - bs\ p, t_2, t_3]$. We now find the parametric equations for b, e and i_j by multiplying TU, giving

$$\begin{aligned} b &= t_3 & (6.5) \\ e &= c_0 bs - bs\ p + c_j t_2 - bs|P|t_3 & (6.6) \\ i_j &= t_2\,, & (6.7) \end{aligned}$$

where all terms on the right (except for the parameters t_i) are known values at run time.

6.3 COMPUTATION PARTITIONING

Given an array reference within a loop, and a surrounding loop nest, it is necessary to find the iterations of the loop nest that access the array. We will show how to do this for each of the four distributions discussed above. In general, we assume an affine expression and bounds on the variables of that expression that describe the elements owned by a process, an affine subscript function $\sigma(I)$ in the index variables of the surrounding loop nest $I = [i_1, i_2, \ldots, i_n]$, and bounds on the loop. Let $i_d = \sigma_d^{-1}(I)$ be an inverse function of the subscript function that returns a value for i_2 given an array element e and bounds on all index variables other than i_d. Our goal is to intersect the elements owned by a process with those accessed by the subscript, and from the resulting solution find the maximum and minimum values of i that are needed to access the elements.

Below, we describe how to find the bounds of the i loop for each dimension of a referenced array. Afterwards we discuss what to do when an array has multiple distributed dimensions, and when there are multiple references in a loop.

6.3.1 BOUNDS OF REPLICATED ARRAY DIMENSIONS

For a replicated distribution, all elements of the array on are each process, and therefore all iterations of the surrounding loop nest must be executed.

6.3.2 BOUNDS OF BLOCK DISTRIBUTED ARRAY DIMENSIONS

For a block distribution, let us first consider the case where σ is an affine subscript expression in a single index variable i_k, i.e., it has the form $e = ci + c_0$, and $i = \frac{e-c_0}{c}$. Given the upper (lower) bound of a on p, $U_{a,p}$ $(L_{a,p})$ then the bounds on i_k are

$$\left\lceil \frac{U_{a,p} - c_0}{c} \right\rceil \leq i \leq \left\lfloor \frac{e - c_0}{c} \right\rfloor .$$

6.3.3 BOUNDS OF CYCLICALLY DISTRIBUTED ARRAY DIMENSIONS

A cyclic reference is somewhat more difficult because the elements are not contiguous, and the stride of the array accesses must be accounted for.

The upper and lower bounds of i_k values for p can be found as above for a block distribution. The elements accessed by the subscript expression are simply the values of σ evaluated over the range given by the upper and lower bounds. The elements that are actually present on the process can be described by the function $e = |P|i' + p, i' \in [0 : |e|]$, where $|e|$ is the number of elements given in Equation 6.1.3. We solve for when the

$$c_1 i + c_0 = |P|i' + p,$$

i.e.,

$$c_1 i - |P|i' = p - c_0.$$

Using techniques from the background section of Chapter 7.1 allows the solution for i to be cast in terms of a parametric equation that can be used to enumerate the loop index values.

6.3.4 BOUNDS OF BLOCK-CYCLICALLY DISTRIBUTED ARRAY DIMENSIONS

The case of block-cyclic is most difficult. What is needed is a loop that traverses the different blocks, a nested loop that traverses the different elements in the blocks, and a value of i in the global iteration space of the loop, and an index into the local array. First, we find the bounds on b, the starting elements of blocks that are within the bounds defined by the block size, array size and iteration space of the loop. From Equation 6.5 we know that $b = t_3$, and therefore the bounds on t_3 are the bounds on b. Next, we find the bounds on t_2, needed to compute the elements e that are accessed within each block. We know that $0 \leq e \leq bs - 1$ from the previous discussion in Section 6.1.4 above. Substituting the right-hand side of the equation for e in Equation 6.5 above we get

$$0 \leq bs - bs\,p + c_j t_2 - bs|P|t_3 \leq bs - 1$$

and solving for t_2 in the inequality gives

$$\left\lceil \frac{-c_0 bs + bs\ p + bs|P|t_3}{c_j} \right\rceil \leq t_2 \leq \left\lceil \frac{-c_0 bs + bs\ p + bs|P|t_3 + bs - 1}{c_j} \right\rceil . \qquad (6.8)$$

In Equation 6.5, $i_j = t_3$, and therefore the inequality above gives the bounds on the i_j. Figure 6.6 shows how the bounds of a loop are transformed to only perform the iterations that access the locally owned elements of a block-cyclically distributed array.

```
for (i = 0; i < U_i; i ++) {
    ... a[σ(i)]...
}
```

(a) An original loop with a block-cyclically distributed array.

```
L_b, U_b as given in Equation 6.3
L_t2, U_t2 as given in Equation 6.8
L_a, U_a are the declared global size of a
for (b = L_t2; b ≤ U_t2; b ++) {
    L_a,b = pbs + L_a + b|P|bs
    e_f = σL_t2 − σ(L_a,b) − c
    for (i = max(L_i, L_t2); i ≤ min(L_i, L_t2); i ++) {
        e_f += c
        ... buffer(b, e_f) ...
}
```

(b) The loop of (a) transformed.

Figure 6.6: An example of determining loops bounds when accessing a block-cyclically distributed array.

6.3.5   SUBSCRIPTS WITH MULTIPLE LOOP INDICES

What do we do if σ is not in a single variable, but contains multiple index variables? Again, assume we are trying to find the bounds on a loop with index i_k. For all loops outside of i_k in the loop nest, the values of their index variables $i_1, i_2, \ldots, i_{k-1}$ are known, and effectively constant, values when an instance of i_k begins to execute.

Thus, the sum of

$$\Sigma_{j=1}^{k-1}(c_j i_j)$$

can be formed and made part of the constant in the bounds formula above. For the loops nested within i_k, i.e., loops with index variables $i_{k+1}, \ldots, i_n$ we know the maximum and minimum values of the index variables. Thus, we can find the extreme values of the integer expression

$$c_{k+1}i_{k+1} + \ldots + c_n + i_n \,, \tag{6.9}$$

as described in Section 7.3. For finding the lower bound of i_k we need the smallest value of $\frac{U_{a,p}-c_0}{c}$, which requires using the lower bound of the expression of Equation 6.9. Similarly, to find the upper bound we need to use the upper bound of the expression of Equation 6.9. Index variables not mentioned in the subscript of the distributed dimension are not constrained by the distribution, and the full range of iterations must be executed.

6.3.6   GENERATING LOOP BOUNDS FOR MULTI-DIMENSIONAL ARRAYS

The techniques described so far in this section compute a set $\mathcal{I}_d$ that contains all iterations in the loop nest surrounding the iteration that must be executed to access the elements owned by a process, and referenced by the subscript expression for a single dimension of an array. In general, arrays will contain multiple distributed dimensions.

In the simple case the different distributed dimensions are indexed by different index variables. Let i_d be the index variable used to index dimension d of the array. Then the constraints imposed on the values of i_d found using the formulae of this section give the necessary bounds on i_d for this reference.

The more complicated case arises when one or more loop indices i_k are used to index several dimensions $i_{d_1}, i_{d_2}, \ldots, i_{d_n}$ of the array. For a value of i_k to be in the iteration space accessing the reference, it must be valid for indexing into each dimension $d_1, d_2, \ldots, d_n$. Therefore, the contribution of this reference to the iteration space of the i_k loop is given by $\mathcal{I}_a = \mathcal{I}_{d_1} \cap \mathcal{I}_{d_1} \cap \ldots, \mathcal{I}_{d_1}$, i.e., the intersection of the iteration space defined by each dimension.

6.3.7   GENERATING LOOP BOUNDS WITH MULTIPLE REFERENCES

Loops with a single reference are exceedingly rare and so it is necessary to combine the iteration spaces defined by each reference into a total iteration space. If some iteration of a loop index i is needed by some reference, it must be included in the iteration space of the loop. Thus, given the sets of iterations $\mathcal{I}_{a_1}, \mathcal{I}_{a_2}, \ldots, \mathcal{I}_{a_r}$ for the r references in the loop, the total loop nest is $\mathcal{I}_{a_1} \cup \mathcal{I}_{a_2} \cup \ldots \cup \mathcal{I}_{a_r}$. Because the iteration sets are unioned, the resulting loop iteration set will likely be too large for at least some references. To protect references from being executed for invalid iterations, a guard must be inserted around each reference a_j that allows the reference to execute only in those iterations where the value of the current loop nest index is in the iteration set for a_i, i.e., when $I \in \mathcal{I}_{a_j}$.

At first this seems horrendously expensive, since in the worst case almost the whole iteration space of the loop will be executed, with expensive guards, implemented as `if` statements, being executed for each reference. Typically, the situation is much better than this, especially in well-written programs. The references within a loop generally access arrays of the same size, and with the same distribution. This leads to similar iteration sets, and loop bounds, being defined for each reference. When bounds do differ, the often differ by one. In this case, *iteration peeling* (see Section

4.1) can be used to pull off the iterations that are not shared by all references, making the majority of the iterations guard free. Even if iteration peeling cannot be used the loop bounds defined by each reference largely overlap, the work done on each process is roughly $1/|P|$ of the total work, and significant parallelism is realized in the execution.

6.4 GENERATING COMMUNICATION

The typical memory model for distributed computation says that a process will execute whatever code is necessary to compute the values of data (typically array elements) that have been distributed onto it, i.e., the owner's computes rule [198]. Computing these values requires accessing data that is distributed onto other processes, and this requires generating message passing communication to get that data.

Generating communication requires two distinct activities. First, the kind of communication needed is determined. This is done by analyzing the dependence relations between writes of owned data, and reads of data that are potentially not owned. In general, any dependence that has a non-equal direction on a parallel loop will need to obtain data from other processes, since the presence of the non-equal direction means that any subset of the iteration space (such as the subset executing in a given process) will need data from outside the set (and therefore executing on another process). Second, the data that must be communicated needs to be determined. Because most MPI communication requires actions by both the sender and the receiver of the data it is necessary for the sending process to know that it must send data, and what data it must send.

There is not an over-arching strategy for generating communication operations in message passing programs, rather there are a collection of heuristics that attempt to discover a write/read pair corresponds to well-known and supported messaging patterns, and if so to generate code that efficiently creates communication that uses that pattern. In this section we will discuss three communication patterns – shift, using send receive; a collective operation; and a wave-front communication pattern – and how they are detected. This will give the general flavor of how communication can be generated by a compiler.

6.4.1 THE SHIFT COMMUNICATION PATTERN

Figure 6.8(a) shows an HPF program with an array A whose columns are block distributed onto four processes. Figure 6.8(b) shows the layout on the processes, with arrow indicating elements of A on one process must be sent to another process when computing values for the elements of A. The compiler computes, or generates code that computes at run time, the elements of A needed by each process P_i. The compiler also computes, or generates code to compute at run time, the values of I_1 and I_2 that define the bounds of the I_1 and I_2 loops on each process. By evaluating the subscript expression of the B reference over those values, the values of B that need to be read can be determined. These values can be intersected (by means of Diophantine equations for general distributions and affine array subscripts) with what is owned by each process, with the result of the intersection being that the left-hand strip of B on each process needs to be sent to the neighboring process to the left.

A simpler technique can also be used for certain subscript patterns. Given two arrays, A and B, that have the same size and distribution, when some column (row) access of B is offset by a constant value $k < 0$ there needs to be a transfer of k columns (rows) of B to the left (upper) process in the grid. If the constant offset is k, then there is a transfer of k columns (rows) to the right (lower) process. In the example of Figure 6.8, $k = -1$. If B is ALIGNed or otherwise has its distribution shifted relative to As distribution, then this shift is used to modify the value of k. Thus, if the HPF alignment distribution ALIGN $B(I_1, I_1 - 1)$ WITH $A(I_1, I_1)$ were used, B would be shifted to the left one row, the value of k would be increased by one, and the final k in the example would now be 0. Similarly, if the (pathological, in this case) alignment of ALIGN $B(I_1, I_1 + 1)$ WITH $A(I_1, I_1)$ were used, the elements of B would be shifted column to the right with respect to B, the final k value would be -2, and two columns of B would need to be communicated to the left process.

Where the communication is placed depends on the dependence structure of the loop. In the program of Figure 6.8(a), the B array is updated inside of the I_1 loop, and the elements updated are elements that will be communicated. Therefore, it is necessary to move the communication outside of this loop.

6.4.2 THE BROADCAST COMMUNICATION PATTERN

Figure 6.9(a) shows an HPF program where the same element (4) of array B is read by every process. Since B is a distributed array, this requires that B(4) be communicated to every process, which is a broadcast communication pattern. The compiler can easily detect this pattern in the program, and by finding the process that owns the constant element can set up the proper broadcast communication operation.

An analogous, but slightly more complicated, example of this is when a column or row of an array is read by every process, and that column or row resides entirely on a single process. Similarly, if an entire array is distributed onto a single process and read by every process, the entire array will need to be broadcast.

A wavefront *communication pattern*   Consider the code of Figure 6.7(a). The parallelism of Figure 6.7(b) can be exposed by generating code as seen in Figure 6.7(c), and exists along the diagonals of the grid (shown with the same shading). The compiler can realize that the loop can be tiled (see Section 5.6) to form the necessary blocks of computation, with the execution order enforced by the communication placed outside of the inner two loops, as shown in Figure 6.7(c).

6.5   DISTRIBUTED MEMORY PROGRAMMING LANGUAGES

Several languages have been developed for targeting distributed memory machines. Nevertheless, the most popular programming model for distributed memory machines is C, C++ or Fortran combined with MPI. Under such a model the compiler's role is to optimize the program as a sequential program—the exploitation of parallelism, data distribution, computation partitioning and generation of communication is entirely the programmer's responsibility. Because MPI is a

```
PROGRAM WAVE
INTEGER A(1000,1000)
PROCESSOR P(N,N)
DISTRIBUTE A(BLOCK,BLOCK) ONTO P
DO I₂ = 1, N, 1
DO I₁ = 1, N, 1
    A(I₁, I₂) = A(I₁, I₂ − 1) + A(I₁ − 1, I₂)
  END DO
END DO
```

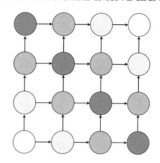

(a) A program with wavefront parallelism.

(b) The iteration space diagram for the program of (a).

```
PROGRAM WAVE
INTEGER A(1000,1000)
PROCESSOR P(N,N)
DISTRIBUTE A(BLOCK,BLOCK) ONTO P
DO II₂ = 1, N, 1
  DO II₁ = 1, N, 1
    if (II₁ > 1) IRECV(A(I₁ − 1, I₂))
    if (II₂ > 1) IRECV(A(I₁, I₂ − 1))
    DO I₂ = 1, N, 1
      DOI₁ = 1, N, 1
        A(I₁, I₂) = A(I₁, I₂ − 1) + A(I₁ − 1, I₂)
      END DO
    END DO
    SEND(up, A(I₁ − 1, I₂))
    SEND(right, A(I₁, I₂ − 1))
  END DO
END DO
```

(c) The program of (a) with the wavefront parallelism of (b) made explicit. Each node in (b) represents a $bs \times bs$ block of computation, i.e., the amount done by one instance of the inner I_1 and I_2 loops.

Figure 6.7: Example of a wavefront pipelining enforced by communication.

library, production compilers have no understanding of the underlying semantics of library calls, and make no effort to optimize code across calls.

We now briefly discuss several other programming languages for distributed computing. These languages fall into the realm of *Parallel Global Address Space* (or PGAS) languages. They allow the programmer to manipulate data using the global indices for distributed arrays. This, in and of itself, is a huge improvement for the programmer over manipulation data using the local address space.

```
PROGRAM SHIFT
INTEGER A(1000,1000)
INTEGER B(1000,1000)
PROCESSOR P(4)
DISTRIBUTE (*,BLOCK) :: A,B ONTO P
DO I₁ = 1, N, 1
    SEND(leftPid, B(I₂, I₁))
    RECV(rightPid, B(I₂, I₁ − 1))
    DO I₂ = 1, N, 1
        A(I₂, I₁) = B(I₂, I₁ − 1)
        B(I₂, I₁ − 1) = . . .
END DO
```

(a) A program with shift communication.

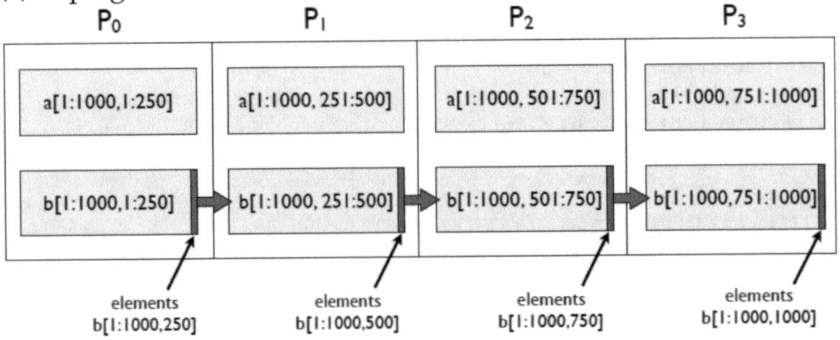

(b) An illustration of the shift communication pattern for the program of (a).

Figure 6.8: An example of a shift communication pattern.

6.5.1   HIGH PERFORMANCE FORTRAN (HPF)

Compilers for High Performance Fortran, and the closely related research languages Fortran90D [35], Vienna Fortran [46], and Fortran D [103], arguably provide the most aggressive support for distributed memory computing. HPF only required the programmer to specify the data distribution of the program. Based on that distribution, the HPF compiler would perform computation partitioning on loops, and automatically generate communication, following the strategies outlined above. It was HPF, and its predecessors, that motivated much of the development and formalization of the work discussed earlier in this chapter, and it was the failure of HPF in the marketplace that led to the rise of Co-Array Fortran (discussed further in Section 6.5.2) and Unified Parallel C (discussed further in 6.5.3).

Novel features of HPF were the ability to declare *processor grids*[1] and *templates*, and to support advanced support of data distribution via declared processor grids, templates, and the *align* primitive. We demonstrate this functionality via an example which is illustrated in Figure 6.10. A 2 × 2 grid

---

[1]These are really grids of *processes*.

```
PROGRAM BCAST
INTEGER A(1000,1000)
INTEGER B(1000)
PROCESSOR P(4)
DISTRIBUTE (*,BLOCK) :: A,B ONTO P
BCAST(B(4))
DO I₁ = 1, N, 1
    DO I₂ = 1, N, 1
    A(I₂, I₁) = B(4)
    END DO
END DO
```

(a) A program with a broadcast communication.

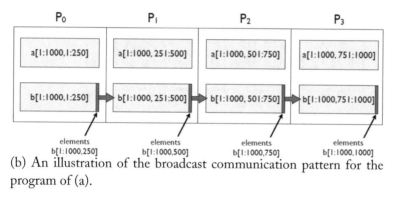

(b) An illustration of the broadcast communication pattern for the program of (a).

Figure 6.9: An example of a broadcast communication pattern.

of processors is declared in $S1$. This is a logical processor grid, in that it will be mapped by the HPF runtime onto whatever number of processors are specified by the programmer.

Statements $S2$ and $S3$ declare a one and two-dimensional array. Remember that Fortran stores arrays in column-major order, and therefore the rightmost subscript corresponds to column elements rather than row elements as it would in C, C++, Java, or most other languages. In statement $S4$, the DISTRIBUTE directive distributes A onto the 4×4 processor grid P using a block distribution. HPF also supported cyclic, block-cyclic and replicated distributions. The ALIGN directive in statement $S5$ places each element B(I) on the same processor that column I of A is placed. The "*" in the second (column) dimension of A says to spread B across all of the columns of the A matrix. In statement $S4$, the DISTRIBUTE directive distributes A onto the 4×4 processor grid P using a block distribution. HPF also supported cyclic, block-cyclic and replicated distributions.

Compiling HPF programs requires many of the analyses and transformations discussed earlier. In the program of Figure 6.11(a) (and its transformed form in Figure 6.11(b)), a flow dependence exists from the write to A[I] to the read of A[I-1]. When compiling for shared memory, three possibilities exist. First, the loop could be left serial. Second, loop fission or distribution could be applied,

```
S1   !HPF$ PROCESSORS P(2,2)
S2   !HPF$ FLOAT A(8,8)
S3   !HPF$ FLOAT B(8)
S4   !HPF$ DISTRIBUTE A(BLOCK,BLOCK) ONTO P
S5   !HPF$ ALIGN B(I) WITH A(I,*)
```

(a) An HPF program using the PROCESSOR grid, alignment, and distribution features of HPF.

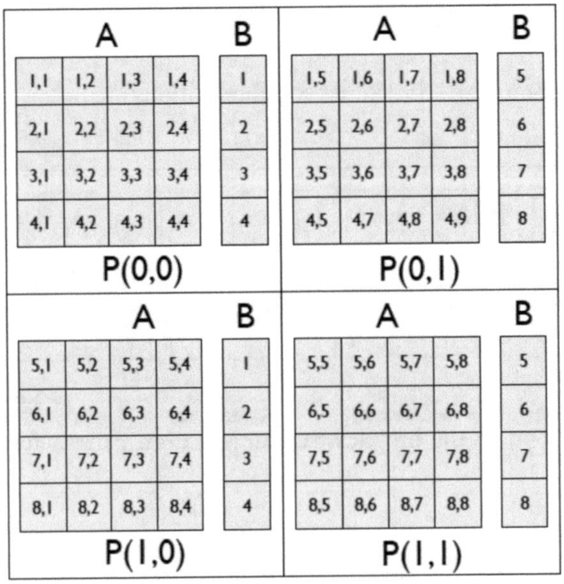

A				B		A				B
1,1	1,2	1,3	1,4	1		1,5	1,6	1,7	1,8	5
2,1	2,2	2,3	2,4	2		2,5	2,6	2,7	2,8	6
3,1	3,2	3,3	3,4	3		3,5	3,6	3,7	3,8	7
4,1	4,2	4,3	4,4	4		4,5	4,7	4,8	4,9	8

P(0,0)   P(0,1)

A				B		A				B
5,1	5,2	5,3	5,4	1		5,5	5,6	5,7	5,8	5
6,1	6,2	6,3	6,4	2		6,5	6,6	6,7	6,8	6
7,1	7,2	7,3	7,4	3		7,5	7,6	7,7	7,8	7
8,1	8,2	8,3	8,4	4		8,5	8,6	8,7	8,8	8

P(1,0)   P(1,1)

(b) The processor grid defined by the program of (a), and how the distribution directives of the program place the data on that processor grid.

Figure 6.10: An example of HPF ALIGN and DISTRIBUTE directives.

and the barrier at the end of the first loop (containing S1) would enforce the dependence. Third, producer/consumer synchronization could be added to enforce the dependence. With a distributed memory target, adding send and receive operations within the loop would enforce the dependences, but because of the great expense of a communication operation, a guard if statement would need to be placed around the send and receive to ensure the send only executed in the last iteration (where data that is written is read by the next processor) and the receive only executes in the first iteration (where data that is read was written by the previous processor). The overhead of the if statements is high enough (both in execution time and in making the flow of control more complicated, potentially hindering other optimizations) that loop fission is applied instead, and the dependence is enforced by the send and receive placed between the two loops rather than by a barrier at the end of the first loop. This code is found in Figure 6.10(b).

```
        DO I=1, N
S1        A[I] = ...
S2        ...= A[I-1]
        END DO
```

```
    ...
    DO I = LB_p, UB_p
S1     A[I] = ...
    END DO
    IF (p < |P|) THEN SEND(A[UB_p], P + 1)
    IF (p > 0) THEN RECV((A[LB_p − 1], P − 1)
    DO I = LB_p, UB_p
S2     ...= A[I-1]
    END DO
```

(a) A loop in an HPF program with a cross-iteration dependence.

(b) Using loop fission and communication to execute the loop of (a) in parallel.

Figure 6.11: An example of transformations necessary to exploit parallelism in an HPF program.

Similarly, the code found in Figure 6.7(b) will, with a good HPF compiler, result in code that uses messages to both communication data and synchronize dependences, as shown in Figure 6.7.

These examples offer some insight into why HPF was even less successful than shared memory auto-parallelizing compilers. First, providing data distribution information gives hints as to how parallel code should be scheduled, but ultimately the problem is the same as confronted by an auto-parallelizing compiler for shared memory. In particular, data dependence analysis is necessary to find parallel loops in a shared memory compiler, and transformations are necessary to either eliminate anti and output dependences, and to enforce flow dependences. In an HPF compiler, dependence analysis is necessary to determine where communication is necessary, what kind of communication is necessary, and what must be communicated. The compiler's job is made easier in some cases by the distributed memory model, for example, cross-iteration (and therefore cross-processor) anti-dependences can be eliminated by a combination of the serial execution within a process, process private storage for all array elements accessed by the process, and by not generating communication that would overwrite the dependent array element on the processor at the sink of a dependence. The distributed memory model makes the compiler's job more difficult because it is necessary to either have exact distance information for cross-iteration flow dependences so that exactly the right data is communicated, or to over-communicate, leading to larger communication overheads. Because message passing communication is typically orders of magnitude more expensive than synchronization in shared memory processors, conservative analyses that lead to unnecessary or inefficient communication can lead to significant slowdowns in program performance. As well, just as interchange and other transformations are necessary when targeting shared memory to form an outer loop that has no dependences and can be parallelized, with HPF dependences must be moved out of the inner loop so that substantial work exists to amortize the cost of the communication. At the end of the day, the technical issues that must be confronted by an HPF compiler are similar, and in many cases fundamentally the same, as those that are confronted by shared memory compilers.

Less technical reasons also led to the lack of wide-spread adoption of HPF. A major issue was the compilation of block-cyclic distributions, or more precisely, the lack of any good method of generating code for block-cyclic distributions when major vendors began work on HPF [36, 89, 98]. Two strategies were followed: (i) vendors chose to simply not support block-cyclic; and (ii) library calls at run time handled the block-cyclic distributions correctly, but inefficiently. The end result of this is that the vendors that supported a subset of the language, usually because of block-cyclic, also decided to not implement some other difficult features. Because all vendors did not make the same choices, different languages were supported by different the compilers. Vendors that chose the runtime approach often focused less on performance and more on completeness—not a winning strategy for a language whose first two initials stand for "high" and "performance". In fairness, it must be said that these vendor focused on performance going forward.

A second major issue was that potential users of HPF had wildly unrealistic expectations of what HPF would be able to accomplish. Compounding this, many HPF compilers gave little or no feedback to their users about where communication was placed, and why. Thus, programmers would compile their application, and get some level of performance, make a few seemingly small changes, and get very different performance. This made it difficult for programmers to work with the compilers to converge on the best performance the compiler could deliver.

6.5.2   CO-ARRAY FORTRAN

Co-Array Fortran is an extension of Fortran 95 and implements an SPMD programming model. It is currently part of the Fortran 2008 standard, released in 2010 [195]. A Co-Array Fortran program is replicated a fixed number of times, and each replication of the program is called an *image*. When the program is replicated, all data associated with the program is, by default, also replicated. Each image has a unique identifier (its *index*) that allows the flow of control within an image to be controlled, as desired, based on the identity of the image.

For useful work to be performed in parallel, data should be distributed among the images so that it can be simultaneously operated on by many processes. Co-arrays give that capability to Co-Array Fortran. Consider the following declaration:

$S1$ REAL, DIMENSION(N)[*] :: A, B
$S2$ A(:) = B(:)[P]

This code does the following things. A statement $S1$ declares two co-arrays named A and B, each with N elements. Thus, each image has a portion of the co-array that contains N elements, and a total of N * *number images* elements are allocated. This statement is very similar to a normal Fortran declaration except for the "[*]" which makes it a co-array.

Statement $S2$ accesses the co-arrays, and assigns all of the values of elements of the B array on image P to the local storage for the co-array Aarray. We note that if every image has the same value of P, every portion of local A array gets the values of the B array from the same image, and thus a broadcast operation has been specified. As the reader may have surmised, the "`:`" notation for A and B arrays specifies that all elements of the array in Fortran 95 syntax, and all elements of the co-array on some image in Co-Array Fortran syntax. In general, for both Co-Array Fortran and Fortran 95, sections of arrays may be specified by triplet notation of the form $[l : u : s]$.

A variety of communication patterns can be specified, using co-arrays and in some cases Fortran 95 *intrinsic* functions[2] Thus, the expression

$$A[P] = B$$

moves the contents of the co-array B on the image executing the statement into the co-array residing on image P. The expression

$$Y[:] = X$$

broadcasts the element values of X on the image executing the statement to every images Y co-array. Note that while "`(:)`" specifies all of the elements of an array, the use of a bracket, e.g., "`[:]`" specifies all processes containing parts of the co-array. As the last example, the expression

$$S = MAXVAL(Y[:])$$

uses the intrinsic `MAXVAL` function perform a max reduction over all elements of Y on all images. If every image executes this statement, every image will get the maximum value contained in some Y in their local scalar variable S.

It is the task of the programmer to map data into the distributed co-array. It is the task of the compiler to recognize remote accesses, generate the message passing code to create these accesses, and to recognize collective communication operations encoded as assignment statements involving co-arrays. Because communication is more explicit than in HPF, the programmer is more aware of the cost of communication operations that are required, and consequently can better manage the communication, and tune the performance of a program, at the source level.

6.5.3 UNIFIED PARALLEL C (UPC)

Unified Parallel C, as implied by the name, is a set of extensions to the C programming language. These extensions include constructs and/or semantics to enable explicit parallelism, support a shared address space across a distributed memory system, perform synchronization and to maintain strict or relaxed consistency models. Like Co-Array Fortran, it is primarily the responsibility of the programmer to tune their program for good performance, and like Co-Array Fortran it gives programmers enough control to achieve good performance while relieving them of the tedium of perform global to local address space calculations when accessing shared data structures. And also like Co-Array Fortran it also assumes an SPMD execution model.

UPC assumes a memory organization as shown in Figure 6.12(a). The memory is physically distributed across $|P|$ processes, and each processes' physical memory is logically partitioned to be

---

[2]Intrinsic functions are Fortran functions that are defined as part of the language specification, and whose semantics are known to the compiler.

in two parts: a globally accessible part treated as shared memory, and a private part only accessible by the process on which the memory resides. Data declared as *shared* is placed into the logically shared portion of the memory, and data declared as *private* is placed into the logically private part of the memory. The default distribution of data across a process is done using a block distribution. By changing the declared block size of data a programmer can achieve a cyclic or block-cyclic distribution. Figure 6.12(b) shows a program that declares three arrays (v, b, c and d) as private, block, cyclic and block-cyclically distributed, respectively. Figure 6.12(c) shows how these arrays are laid out in memory.

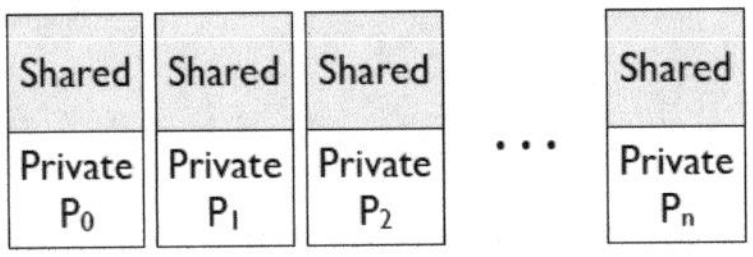

(a) The organization of memory by the UPC compiler and run time.

```
float v[100]
shared [ceil(100/THREADS)] float b[100]
shared float c[100]
shared [2] float d[100]
```

(b) Declarations of four variables with different distribution attributes.
v is private, b is block distributed, c is cyclical distributed, and d is block-cyclically distributed with a block size of 2.

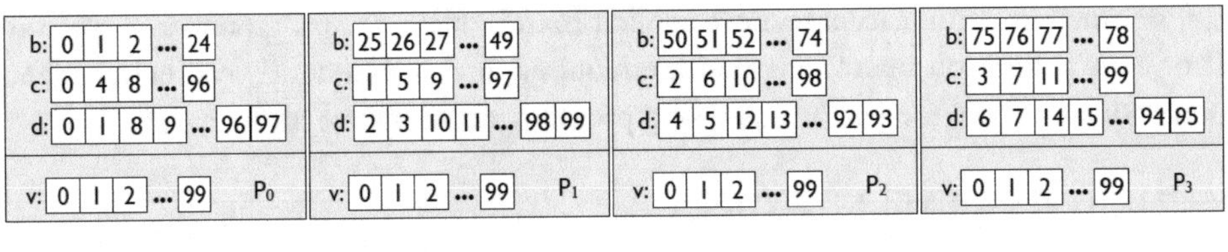

(c) The layout of the arrays a, b, c and d above.

Figure 6.12: Memory organization and data distribution in UPC.

Since UPC extends C, support for pointers into distributed data is mandatory. As can be seen in Figure 6.12(b), the location of data, and the distribution of data that resides in shared memory is expressed as a type. Therefore, UPC supports pointers to shared data.

UPC pointers can reside in either private or shared memory, and can point to either private or shared memory. A pointer in private memory that points to private memory provides a pointer that can only be changed by one UPC thread, and that points to data in that threads shared memory. An example of this is pointer p in Figure 6.13(a). A pointer in private memory that points to shared data

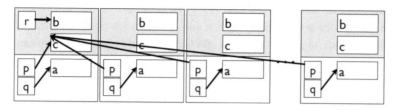

(a) An example of shared and private pointers that point-to shared and distributed data.

```
float * q;
shared float *p;
shared float shared *r;

q = &a;
p = &c;
r = &b;
```

```
shared [2] float a[10], *p;
shared [3] float *q;
...
p = a; p+ = 4;
q = a; q+ = 4;
```

(b) Declarations of shared and private pointers to access the arrays.

(c) A program doing pointer arithmetic on pointer to shared arrays.

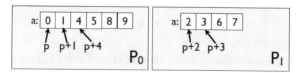

(d) A graphical illustration of the (probably correct) pointer arithmetic on p in the program fragment of (b).

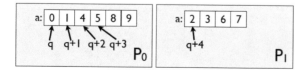

(e) A graphical illustration of the (probably incorrect) pointer arithmetic on q in the program fragment of (b).

Figure 6.13: An example of UPC pointers and pointer declarations.

allows a thread to have a pointer that cannot be changed by other threads and from which shared memory values can be accessed. An example of this is pointer q in Figure 6.13(a). A pointer in shared memory pointing to shared storage in shared memory provides a pointer than can be read and written by all threads, and that points to data accessible by all threads. An example of this is pointer r in Figure 6.13(a). The last combination of pointer location and the location of pointed-to storage is a pointer in shared memory that points to private memory. This pointer combination is prohibited by UPC, since it would allow the address of private data to be accessed by all threads. Moreover, since global pointers contain header information that provide information about which thread contains the pointed-to data, and pointers to private storage do not, dereferencing such a pointer would likely point into the dereferencing threads private storage, even if the pointer value was set by another thread. Figures 6.13(c) and 6.13(d) show pointers to the arrays declared in Figure 6.12(b).

Figures 6.13(d) and 6.13(e) show pointers to data with different block sizes, and show that like sequential C, pointer arithmetic honors type information. In UPC, the block size is part of the type, and so the block information associated with a pointer, rather than with the pointed-to object, will affect the pointer arithmetic.

UPC also supports both relaxed and strict (sequentially consistent) memory models. Critical cycle analysis for SPMD programs, as described in Section 2.6, is used constrain compiler optimizations, and to guide the insertion of fence instructions to constrain reordering of instructions by the process.

In addition to the memory model analysis just mentioned, the UPC compiler performs a mapping from global to local index spaces for distributed arrays. It also must generate communication when accessing non-locally stored data, and when accessing locally stored (but perhaps shared or private) data it should either determine statically or at run time that the data is local and use a load or store rather than message passing to access the code.

Both UPC and Co-array Fortran have enjoyed more success that HPF because they both give the responsibility for performance to the programmer, provide a programming model that allows the programmer to gain some insight into the performance of the program being written, and relieve the programmer of the tedium of managing data distributions and the details of communication.

CHAPTER 7

Solving Diophantine equations

Several topics discussed in this lecture have required solving *Diophantine equations* (e.g., Sections 2.3.2 and 6.1). Diophantine equations are equations whose domain, range and coefficients are integers and whose form is that of an indeterminate (i.e., it has infinite solutions) polynomial. The equations we are interested in are polynomials of degree one in possibly many variables, which are often referred to as affine equations. This chapter may be skipped, or quickly scanned, by most readers, but is provided for the reader desiring a deeper understanding of the mechanics behind solving Diophantine equations.

7.1 SOLVING SINGLE DIOPHANTINE EQUATIONS

Let the Diophantine equation we wish to solve be in the variables $i_1, i_2, \ldots, i_n$, and of the form:

$$c_0 = a_1 i_1 + a_2 i_2 + \ldots + a_n i_n . \tag{7.1}$$

We can form a coefficient matrix A of the equation:

$$A = \begin{bmatrix} a_1 \\ a_2 \\ \ldots \\ a_n \end{bmatrix} ,$$

where the j'th row of A is the coefficient for the j'th variable in the system. The left-hand side of the system, C, can be similarly represented. Because our system only has one equation, this is simply:

$$C = \begin{bmatrix} c_0 \end{bmatrix} .$$

To solve the system of equations, we first form the matrix IA, where I is the identity matrix. This gives the matrix

$$IA = \begin{bmatrix} 1 & 0 & 0 & \ldots & 0 & a_1 \\ 0 & 1 & 0 & \ldots & 0 & a_2 \\ \ldots & & & & & \\ 0 & 0 & 0 & \ldots & 1 & a_n \end{bmatrix} .$$

An elimination algorithm, reminiscent of Gaussian elimination but using only integer operations, can be applied to the matrix to obtain the matrix UD:

$$UD = \begin{bmatrix} u_{1,1} & u_{2,1} & \cdots & u_{n,1} & d_1 \\ u_{1,2} & u_{2,2} & \cdots & u_{n,2} & 0 \\ \cdots & & & & \\ u_{1,n} & u_{2,n} & \cdots & u_{n,n} & 0 \end{bmatrix}.$$

The resulting U matrix is an $n \times n$ matrix, and the resulting D matrix contains the greatest common divisor (or *gcd*) of the coefficients for the equation as its first element.

Let $T = [t_1, t_2, \ldots, t_n]$ be a vector of n parameters. Then $TD = C$, i.e., in our equation, $c_0 = d_1 t_1$, and by simple algebra, $t_1 = d/d_1$. The equation has a solution if and only if d_1 evenly divides c, i.e., if the gcd of the coefficients evenly divides the constant term. If a solution does exist, then

$$T = \left[t_1 = \frac{c}{d_1}, t_2, \ldots, t_n \right]$$

and the relationship

$$[i_1, i_2, \ldots, i_n] = TU$$

holds. Stated more verbosely,

$$
\begin{aligned}
i_1 &= \frac{c}{d_1} u_{1,1} + t_2 u_{1,2} + \ldots + t_n u_{1,n} \\
i_2 &= \frac{c}{d_1} u_{2,1} + t_2 u_{2,2} + \ldots + t_n u_{2,n} \\
&\cdots \\
i_n &= \frac{c}{d_1} u_{n,1} + t_2 u_{n,2} + \ldots + t_n u_{n,n}.
\end{aligned}
$$

By substituting the solutions for t_1 and arbitrary integer values for t_k, $1 < k \leq n$ into the system of Equation 7.2, all solutions to Equation 7.1 can be found.

7.2 SOLVING MULTIPLE DIOPHANTINE EQUATIONS

It is sometimes desirable to simultaneously solve a system of Diophantine equations. A general system of Diophantine equations can be expressed as:

$$
\begin{aligned}
c_0 &= a_{1,1} i_1 + a_{1,2} i_2 + \ldots + a_{1,n} i_n \\
c_0 &= a_{2,1} i_1 + a_{2,2} i_2 + \ldots + a_{2,n} i_n \\
&\cdots \\
c_0 &= a_{m,1} i_1 + a_{m,2} i_2 + \ldots + a_{m,n} i_n.
\end{aligned}
$$

This set of equations can be represented in matrix form as:

$$\begin{bmatrix} a_{1,1} & a_{1,2} & \cdots & a_{1,n} \\ a_{2,1} & a_{2,2} & \cdots & a_{2,n} \\ & & \cdots & \\ a_{m,1} & a_{m,2} & \cdots & a_{m,n} \end{bmatrix} \times \begin{bmatrix} i_1 \\ i_2 \\ \cdots \\ i_n \end{bmatrix} = \begin{bmatrix} c_1 \\ c_2 \\ \cdots \\ c_n \end{bmatrix},$$

or, more generally, $iA = C$ and the unique solution to this equation can be found as $i = CA^{-1}$.

The goal in solving such system is the same as when using Gaussian elimination—we wish to reduce the array and find an upper triangular array that can be used to solve for each of the n variables in turn. However, because we are working with Diophantine equations, and we want them to remain Diophantine equations throughout the process, we avoid the use of division in the elimination process. Two types of operations will be performed on the matrix, interchange of two rows r_p, r_q and replacement of a row r_p by $r_p - kr_q$. By performing these operations repeatedly, the matrix is reduced to an upper triangular form.

We now describe how to do this systematically, additional details can be found in [24]. Consider the $n \times m$ matrix of integers, $\mathbf{A}$, above. It can be augmented with an $n \times n$ identity matrix $\mathbf{I}$, yielding the matrix

$$\begin{bmatrix} a_{1,1} & a_{2,1} & \cdots & a_{m,1} & 1 & 0 & \cdots & 0 \\ a_{1,2} & a_{2,2} & \cdots & a_{m,2} & 0 & 1 & \cdots & 0 \\ & & \cdots & & & & \cdots & 0 \\ a_{1,n} & a_{2,n} & \cdots & a_{m,n} & 0 & 0 & \cdots & 1 \end{bmatrix}.$$

We repeatedly apply the operations of Section 7.1 to eliminate various elements of the $\mathbf{A}$ matrix until the upper triangular form is achieved. At this point we have an $n \times m$ upper triangular transformed matrix $\mathbf{A}$ and a $n \times n$ Unimodular matrix $\mathbf{U}$ that is the result of applying the operations forming the upper triangular $\mathbf{A}$ matrix to the $\mathbf{I}$ matrix. As in the previous section, $x = TU$. The the parametric solutions of the system of Diophantine equations can be found by multiplying the vector $[c', t_2, t_3, \ldots, t_n]$ times $\mathbf{U}$. An example is given in Figure 7.1.

7.3 EXTREME VALUES OF INTEGER FUNCTIONS

It is often the case that it is desirable to find the minimum and maximum values that an affine expression can assume. In this section we will show how to find the value of the expression

$$a_1 i_1 + a_2 i_2 + \ldots + a_n i_n - a_0 ,$$

where, as above, the a_k are integer constants and the i_k are integer variables. For each i_k we assume there are integer bounds such that $L_{i_k} \leq i_k \leq U_{i_k}$. For additional details, see [24].

We first provide two definitions.

Definition 7.1 The *positive* part of a number a, denoted a^+, is a when $a \geq 0$, and is 0 otherwise.

$$2 * i_1 + 2 * i_2 + 3 * i_3 = 2$$

$$\begin{aligned} a_1 &= 2 \\ a_2 &= 2 \\ a_3 &= 3 \\ c_1 &= 2 \end{aligned}$$

$$\begin{bmatrix} 1 & 0 & 0 & 2 \\ 0 & 1 & 0 & 2 \\ 0 & 0 & 1 & 3 \end{bmatrix}$$

(a) A Diophantine equation to be solved.

(b) A mapping of coefficients and constants to the notation for solving Diophantine equations.

(c) The matrix $\mathbf{I\,A}$.

$$\begin{bmatrix} 1 & 0 & 0 & 2 \\ 0 & 0 & 1 & 3 \\ 0 & 1 & 0 & 2 \end{bmatrix}$$

$$\begin{bmatrix} 1 & 0 & 0 & 2 \\ 0 & -1 & 1 & 1 \\ 0 & 1 & 0 & 2 \end{bmatrix}$$

$$\begin{bmatrix} 1 & 0 & 0 & 2 \\ 0 & 1 & 0 & 2 \\ 0 & -1 & 1 & 1 \end{bmatrix}$$

(d) The matrix of (c) after switching rows 2 and 3.

(e) The matrix of (d) after subtracting row 3 from row 2.

(f) The matrix of (e) after switching rows 2 and 3.

$$\begin{bmatrix} 1 & 0 & 0 & 2 \\ 0 & 3 & -2 & 0 \\ 0 & -1 & 1 & 1 \end{bmatrix}$$

$$\begin{bmatrix} 1 & 0 & 0 & 2 \\ 0 & -1 & 1 & 1 \\ 0 & 3 & -2 & 0 \end{bmatrix}$$

$$\begin{bmatrix} 1 & 2 & -1 & 0 \\ 0 & -1 & 1 & 1 \\ 0 & 3 & -2 & 0 \end{bmatrix}$$

(g) The matrix of (f) after twice subtracting row three from row 2.

(h) The matrix of (g) after switching rows 2 and 3.

(i) The matrix of (h) after twice subtracting row 2 from row 1.

$$\begin{bmatrix} 0 & -1 & 1 & 1 \\ 1 & 2 & -1 & 0 \\ 0 & 3 & -2 & 0 \end{bmatrix}$$

$$(i_1, i_2, i_3) = (2, t_2, t_3)\mathbf{U}$$

$$\begin{aligned} i_1 &= t_2 \\ i_2 &= 2t_2 + 3t_3 - 2 \\ i_3 &= -t_2 + -t_3 + 2 \end{aligned}$$

(j) The matrix of (g) after switching rows 1 and 2.

(k) i_1, i_2 and i_3 expressed in terms of parameters t_2, t_3 and $\mathbf{U}$.

(l) Equations for i_1, i_2 and i_3 expressed in terms of $t_1 = c/d$, t_2 and t_3.

Figure 7.1: An example of solving a Diophantine equation and expressing its solution in terms of parametric terms.

Definition 7.2   The *negative* part of a number a, denoted a^-, is $|a|$ when $a \geq 0$, and is 0 otherwise.

Thus, the negative and positive parts of a number return the magnitude of the number when it is a negative, and positive number, respectively. Using these concepts, we can succinctly express the extreme values of the expression. The lower bound of the expression can be

$$L_E = a_0 + \Sigma_{k=1}^n L_{i_k} a_k^+ - U_{i_k} a_k^- \tag{7.2}$$

and

$$U_E = a_0 + \Sigma_{k=1}^n U_{i_k} a_k^+ - L_{i_k} a_k^- . \tag{7.3}$$

Intuitively, when computing the lower (upper) bound we want to minimize (maximize) the value of each term. To minimize a term when the coefficient a_k is positive requires using the lowest value i_k can take on, and when the coefficient a_k is negative using the largest value of i_k, and when a_k is positive using the smallest value of i_k. The maximum value is done in a similar, and symmetric, way.

CHAPTER 8

A guide to further reading

Because of the tutorial focus of this book I have focused on giving relatively high-level sources, where possible, to supplement the material.

8.1 COMPILER FUNDAMENTALS

There are a variety of compiler textbooks available, and most of these cover parsing, code generation, and optimizations and analyses for sequential machines. The so called *Dragon Book* (because of its cover) is a classic in the field, but fairly technical. Two other, more accessible books are Fischer and Cytron's *Crafting a Compiler* [77] and Torczon and Cooper's *Engineering a Compiler* [57]. For advanced coverage, Muchnick's *Advanced Compiler Design & Implementation* [161] surveys a range of advanced topics. Because of the breadth of the field, all compiler books are somewhat flawed, and the reader with finite time and resources is suggested to examine two or three from a library before settling on one.

The development of the Static Single Assignment (SSA) form in the 1980s had a significant impact on the design of intermediate representations for scalar compilers. A good early work on the program dependence graph, a precursor to the SSA form, is in Ferrante, Ottenstein and Warren paper [74]. The SSA form itself is discussed in detail in [62], and detailed development of an algorithm for using the SSA form to perform constant propagation is given in [240].

Padua's 2000 article on *The Fortran I Compiler* [167] serves as a fascinating look into the first compiler, and gives a good perspective on how the initial technology has been replaced by more advanced and formalized methods.

An alternative approach to developing parallel programs is write a sequential program on top of a parallel library, with the library providing the parallelism. This has been used successfully for decades in the more limited domain of databases. Two projects along these lines are Rauchwerger's STAPL [192, 217, 226] and Intel's Thread Building Blocks [229].

8.2 DEPENDENCE ANALYSIS, DEPENDENCE GRAPHS AND ALIAS ANALYSIS

Good overviews of dependence analysis techniques can be found in the tutorials by Padua and Wolfe [169], Wolfe [249, 251] and Banerjee et al. [23]. Modern dependence analysis grew out of Utpal Banerjee's PhD thesis at Illinois in the late 1970s. Two books give an in-depth explanation of classical Banerjee-Wolfe dependence analysis [22, 24] as well as explaining the necessary number theory and Unimodular operations on matrices. The textbooks by Kennedy and Allen [116] and Wolfe [250] also provide a good discussion of this topic. A comparative overview of dependence testing techniques can be found at [104].

In the years following the development of classical Banerjee-Wolfe dependence analysis, an number of different approaches have been taken to make the results less conservative. Early variants were the *Power Test* [252], which used Fourier-Motzkin elimination to decide if references are dependent, and the i-test [125]. Another variant is the Range Test [31], which determines that the range of values taken on by subscripts in potentially dependent array accesses do not overlap, and thus cannot take on the same value. Even more precise tests are the Omega test [181, 183, 184] and the Polyhedral and Polytope models [26, 33, 72, 73, 177, 178]. Both of these have significantly higher theoretical overheads than Banjerjee-Wolfe analysis, but are practical for use in a compiler.

Readers with an interest in analyzing explicitly parallel programs should first read Adve and Gharachorloo's tutorial on shared memory consistency models [2]. Many of the issues in compiling for different memory models can be understood using the techniques of [209]. Problems in the original Java memory model are discussed in [185], and the current Java memory model, released with Java 5, is discussed in Manson, Pugh and Adve [147]. Finally, Boehm and Adve [32] did fundamental work in developing a memory model for C++. The two main compiler efforts targeting the compilation of explicitly parallel programs while preserving sequential consistency are Yelick's Titanium [127, 128] and UPC [224], and the Pensieve project [141, 225].

8.3   PROGRAM PARALLELIZATION

For a good overview of parallelization, the sources mentioned earlier [23, 169, 244, 251] should be consulted. Other good overviews include [132]. Vectorization is discussed in the above works, and [9, 10, 40, 247]. Much of this work was inspired by the early CCD and Cray machines. More recently, the introduction of short vector instructions in the Intel and PowerPC processors has led to some recent efforts in this area, including [70, 196, 214], as has compiling for graphics processors being used as accelerators [38, 142, 149, 205, 227] and the Cuda [60] language compilers.

Related to dependence analysis and automatic parallelization, Rauchwerger, et al. did work in hybrid parallelization, where some information is developed at compile time but the final decision on whether a dependence exists and the loop can be parallelized is made at run time [202, 203].

Good overviews of producer/consumer synchronization can be found in [157, 246]. Much of the early research on producer/consumer synchronization [144, 156, 223], came from the Cedar Project [131], which implemented flexible synchronization mechanisms. The Rice compiler effort, under Ken Kennedy, also studied the problem of compiler generation of synchronization [42]. In [49], the effects of synchronization and the granularity of parallelism enabled by synchronization are discussed. With multi-core processors there has been some renewed interest in this topic, with some recent work being described in [115, 206, 257].

The parallelization of recursive and divide and conquer algorithms and while loops are related and is an intrinsically more difficult problem than the parallelization of loops, and in the domain of scientific computing these programs have been less common than programs with loop-based computation. Some readings in this area are [56, 91, 108, 109, 193, 201].

A good high-level discussion of software pipelining can be found in [20] and [5, 113] gives a comparison of some of the major techniques. Early work derived from compiling for VLIW machines [4, 68, 80, 81, 138, 171, 191, 199, 232].

8.4   TRANSFORMATIONS TO MODIFY AND ELIMINATE DEPENDENCES

A good overview of loop skewing can be found in [244, 250]. In [21, 140, 243], they discuss loop skewing in the context of Unimodular transformations. An example of a recent work on skewing is [112], which also discusses skewing of multiple loops.

Induction variable substitution is discussed in [3, 96, 176, 248, 253]. An discussion of induction variable analysis in recursive programs using abstract interpretation can be found in [15]

One of the earliest descriptions of the use of forward substitution can be found in [133]. An overview of the algorithm can be found in [169], a description of its implementation within a symbolic analysis framework is described in [94]. Descriptions of the use of forward substitution in parallelizing compilers are given by [30, 95, 210].

The overview works, e.g. [169, 250], give a good overview of scalar expansion and privatization. The transformation is discussed in the context of distributed memory machines by Banerjee, et al. in [170]. The seminal works on array privatization are by Padua and Tu [233, 234].

8.5   REDUCTION RECOGNITION

Kuck [130] provided one of the earliest discussions of the principles of reduction and recurrence recognition and optimization. Other early works on this topic include [78, 114, 175, 176] In [79] the closely related topic of generating parallel prefix programs from recurrences and reductions on data is discussed.

8.6   TRANSFORMATION OF ITERATIVE AND RECURSIVE CONSTRUCTS

Loop fusion and unrolling are closely related to the *unroll-and-jam* transformation [41, 44, 45] that operates on nested loop. Unroll and jam has been applied to many problems, of related to locality, as discussed in [150].

8.7   TILING

An example of an early dependence-based transformation to improve locality can be found in [83]. True tiling is developed in [242, 245]. More advance algorithms were developed by McKinley [150] and Pingali [121]. Related in function, but not in strategy, are various recursive layouts. Gustavson et al. [71, 93] and Chaterjee et al. [48] were pioneers in this. Because of the difficulty of analytically determining the precise tile sizes to best manage a complex memory hierarchy involving multi-level caches and translation look-aside buffers, automatic tuning tools that perform and intelligent search over a space of solutions, using run time profiling, to determine an optimal blocking for a given architecture. ATLAS [18, 254] (for linear algebra) and FFTW [76, 82] (for fast Fourier transforms).

8.8 COMPILING FOR DISTRIBUTED MEMORY MACHINES

Much of the early work in data distribution, communication generation, and so forth, occurred in the context of HPF [117, 123, 124]. Early versions of HPF targeted a variety of message passing systems, but currently the MPI [160, 213] library is used almost exclusively[1]. HPF grew out of three earlier projects – Fortran90D [35] project at Syracuse, Vienna Fortran [46] at the University of Vienna, and Fortran D [103] at Syracuse. As mentioned earlier, the block-cyclic distribution was a difficult problem for early HPF and distributed memory compilers, and Chatterjee's survey paper [239] gives pointers to many of the proposed solutions.

The earliest, and foundational work on data distributions is Koelbel's [122] in his discussion of *local functions*. Much of the later research was focused on relieving the programmer of the task of doing data distribution, and work by Gupta and Banerjee [88] (in the context of the Paradigm Compiler), and Kremer and Kennedy [118]. More advanced generalizations beyond what we have discussed can be found in [86, 235].

Key works in generating communication are again Koelbel's [122], Gupta and Banerjee [88] and FortranD [103], as well as in the Fortran 90D [35] and Vienna Fortran [46] projects.

In [89], an overview is provided of the IBM research prototype that led to the *xlHPF* compiler. This book gives a good overview of the interaction between data distribution, communication generation, computation partitioning, and optimization of communication in the interface between the code and library. Two other important industrial compilers are the DEC compiler [98] and the PGI pghpf compiler [36].

Two of the more successful languages and associated compilers targeting distributed memory computers are Co-Array Fortran (CAF) [54, 163] and Unified Parallel C [43, 237]. A comparison of these two global address space languages is given in [55]. Co-Array Fortran is included in the Fortran 2008 standard approved in 2010 [195].

There are several implementations of both of these languages. The Rice Co-Array Fortran compiler is described in [66], and differs from the standard version by introducing additional features, such as processor sub-setting.

---

[1]MPI is also used in almost all manually written distributed memory programs.

8.9   CURRENT AND FUTURE DIRECTIONS IN PARALLELIZING COMPILERS

There have been two main directions in parallelizing compiler technology in the last 10 or 15 years. These directions have been motivated by a desire to overcome the limitations of purely static compiler analysis. These limitations arise when attempting to parallelize non-loop code; loops with non-affine accesses, including pointer-based accesses; or loop-based and non-loop code with dependences that occur in a few executions of the code.

The first direction to overcome the limitations of the current analyses, and which already has been discussed in the this work, in the use of languages that provide parallelization hints in one form or another to the compiler. These provide information to the compiler above and beyond what it can ascertain via a static analysis. The most widely used of these is probably is OpenMP. The main limitation of this approach is that the languages tend to provide better support for loop and array-based programs and regular algorithms.

The second direction is the use of *speculation*, with the earliest compiler oriented work being that of [56, 63, 64, 193, 194, 218]. Speculative techniques optimistically parallelize program constructs, thus dependences *may* exist across code executing in parallel, and therefore some kind of check must be done at runtime to (i) ensure that the code executing in parallel really is independent and not dependences exist, and (ii) to roll back the execution of one or more of the code instances when a dependence is found, and to re-execute them correctly. Speculation is discussed briefly in Section 3.6.2. The role of the compiler in the current and recent work in speculation is to determine what code should be speculatively parallelized, to transform the code to make the speculative execution more efficient, and to insert the necessary library calls, etc., to enable the speculative execution. Another form of speculation is based on value prediction, where the value that will be communicated along a dependence is predicted (and speculated on), allowing the code at the sink of the dependence to execute before the value is actually computed. If the value is mis-predicted, the dependent code must be re-run.

Speculative approaches need to perform several tasks. As identified in [52], these include:

1. Identifying data references that must be speculated on. These are dependences that are either unlikely to actually exist at runtime, or where the source and sink of the dependence are unlikely to execute at the same time.

2. Maintain information about speculated data locations that have been accessed (e.g. [84]).

3. Schedule threads on which speculation occurs. Like all scheduling strategies it is necessary to worry about load balance. A major contribution of [52] is the use of a sliding window to enable good load balance and the amount of "roll back" necessary to achieve a correct execution.

4. Committing data when it is safe to do so. If a chunk of code has executed with no conflicts on speculated data, the results of that chunk can be committed to memory.

5. Squashing a restarting threads when conflicts on speculated data are detected. Because we are executing speculatively, it will sometimes be true that a dependence does exist, and is violated by the speculated execution. This must be detected, the violating threads terminated, and execution restarted from a correct point.

These basic actions can either be done in software, or more efficiently using special hardware [53, 99]. We note that the first hardware support for transactional memory will likely be in IBM's Blue Gene/Q processor [97]. In [53], a description of a processor that enables thread level speculation, and supporting software, is provided. Unfortunately, software transaction memory systems (STMs) have relatively high costs for the above functionality, and often experienced degraded performance, relative to a sequential version of the program, on a small number of processors. One of the better STMs is STMlite [151] which is designed to use profiling information to facilitate automatic parallelization.

A study of the maximum performance that can be gained by speculative execution given zero-overhead synchronization, etc., and the effects of loop and function level parallelism is provided in [165]. They conclude that function level speculation is best, although much later work mentioned in this section suggests that both coarse and fine grained (loop iterations and other small sections of code) can be effectively exploited. In [28], some issues surrounding the scheduling and selection of code to be executed speculatively, with an emphasis on when to spawn speculative threads, is given. Gonzalez et al. [148] also discuss scheduling strategies utilizing profiling information to increase the ability of the compiler to make better decisions. In [67], a cost-model based framework for selecting and scheduling speculative threads is discussed, and [231] discusses the use of profiling information and machine learning techniques to develop better mappings of speculative threads to different hardware. The POSH compiler [146] uses profiling information and program structures (e.g., loops and functions) to partition programs, and shows that much of the performance improvement comes from overlapping work and prefetching.

Another strategy is to use *program slices* [241] (see Section 2.3.1 to execute a fast path through the program to compute essential values and allow the rapid fulfillment of dependences. This technique, in conjunction with profiling, is explored in [186]. This technique is used to pre-compute addresses (e.g., when following links in a pointer-based data structure) and branch outcomes in [258], and in [259] it is shown that the slice does not have to be correct since if the outcome is incorrect the underlying speculative mechanism allows the mis-calculation to be caught and a correct execution to be done. This observation allows the slice to be highly optimized, which is important since the time it takes the slice to execute affects the rate at which new work can be spawned, and therefore the potential parallelism.

Speculative support for *decoupled software pipelining* (discussed in Section 3.2) is one major approach. As discussed in Section 3.2, decoupled software pipelining executes SCCs across different threads, depends on their being multiple SCCs to form the different pipeline stages, and the performance of the resulting program is gated by the maximally sized SCC. By speculating on some dependences in the loop, more, and better balanced in size SCCs can be formed. Decoupled software pipelining and various optimizations using speculation are discussed in [37, 166, 188, 189, 190, 238].

A significant improvement in the performance of speculatively executed programs can be achieved by making use of high-level semantic information about the operations being executed speculatively. The Galois project [135, 136, 137] makes use of commutativity information as well as high level iterators and novel locking techniques.

Traditional transformations, along with new techniques, can help with exposing regions of code that are amenable to speculative parallelization [256].

Some early work in the use of value prediction and communication of values within the processor to enable thread level speculation is described in [219, 220, 255]. In [65], the use of value prediction as well as speculation on dependences is used to allow more parallelism in the program to be exploited.

Bibliography

[1] W. A. Abu-Sufah and A. D. Malony. Vector processing on the Alliant FX/8 multiprocessor. In *Proceedings of the International Conference on Parallel Processing*, pages 559–566, 1986. Cited on page(s) 58

[2] S. V. Adve and K. Gharachorloo. Shared memory consistency models: A tutorial. *IEEE Computer*, 29(12):66–76, 1996. DOI: 10.1109/2.546611 Cited on page(s) 126

[3] A. V. Aho, M. Lam, R. Sethi, and J. D. Ullman. *Compilers: principles, techniques, and tools*. Addison-Wesley Longman Publishing Co., Inc., Boston, MA, USA, 2006. Cited on page(s) 11, 19, 127

[4] A. Aiken and A. Nicolau. Perfect pipelining: A new loop parallelization technique. Technical report, Cornell University, 1987. Cited on page(s) 127

[5] V. H. Allan, R. B. Jones, R. M. Lee, and S. J. Allan. Software pipelining. *ACM Comput. Surv.*, 27:367–432, September 1995. DOI: 10.1145/212094.212131 Cited on page(s) 127

[6] F. E. Allen, M. G. Burke, P. Charles, R. Cytron, and J. Ferrante. An overview of the PTRAN analysis system for multiprocessing. *J. of Parallel Distributed Computing*, 5(5):617–640, 1988. DOI: 10.1016/0743-7315(88)90015-9 Cited on page(s) 1

[7] J. R. Allen, K. Kennedy, C. Porterfield, and J. Warren. Conversion of control dependence to data dependence. In *Proceedings of the 10th ACM SIGACT-SIGPLAN Symposium on Principles of programming languages*, pages 177–189, New York, NY, USA, 1983. ACM. DOI: 10.1145/567067.567085 Cited on page(s) 39

[8] R. Allen, D. Bäumgartner, K. Kennedy, and A. Porterfield. Ptool : A semi-automatic parallel programming assistant. In *1986 International Conference on Parallel Programming*, pages 164–170, 1986. Cited on page(s) 1

[9] R. Allen and S. Johnson. Compiling C for vectorization, parallelization, and inline expansion. In *Proceedings of the ACM SIGPLAN 1988 conference on Programming Language design and Implementation*, pages 241–249, New York, NY, USA, 1988. ACM. DOI: 10.1145/53990.54014 Cited on page(s) 126

[10] R. Allen and K. Kennedy. Automatic translation of Fortran programs to vector form. *ACM Trans. Program. Lang. Syst.*, 9(4):491–542, 1987. DOI: 10.1145/29873.29875 Cited on page(s) 126

[11] AltiVec technologies. `http://www.freescale.com/webapp/sps/site/overview.jsp?code=DRPPCALTVC`. Last accessed January 5, 2012. Cited on page(s) 7

[12] Unrolling AltiVec, Part 1: Introducing the PowerPC SIMD unit. `http://www.ibm.com/developerworks/power/library/pa-unrollav1/`. Last accessed January 5, 2012. Cited on page(s) 7

[13] Graphics cards from AMD. `http://sites.amd.com/us/game/products/graphics/Pages/graphics.aspx?lid=Gaming_Graphics&lpos=HP_bottom_bucket`. Last accessed January 5, 2012. Cited on page(s) 7

[14] Multi-core processing with amd. `http://www.amd.com/us/products/technologies/multi-core-processing/Pages/multi-core-processing.aspx`. Last accessed January 5, 2012. Cited on page(s) 9

[15] Z. Ammarguellat and W. L. Harrison, III. Automatic recognition of induction variables and recurrence relations by abstract interpretation. In *Proceedings of the ACM SIGPLAN 1990 conference on Programming language design and implementation*, volume 25, pages 283–295, New York, NY, USA, June 1990. ACM. DOI: 10.1145/93542.93583 Cited on page(s) 127

[16] Antlr v3. `http://antlr.org/`. Last accessed January 5, 2012. Cited on page(s) 11

[17] A. W. Appel and J. Palsberg. *Modern Compiler Implementation in Java*. Cambridge University Press, New York, NY, USA, 2nd edition, 2003. Cited on page(s) 11, 19

[18] Automatically Tuned Linear Algebra Software (ATLAS). Downloaded from `http://math-atlas.sourceforge.net/`. Last accessed January 5, 2012. Cited on page(s) 128

[19] M. Bach, M. Charney, R. Cohn, T. Devor, E. Demikovsky, K. Hazelwood, A. Jaleel, C.-K. Luk, G. Lyons, H. Patil, and A. Tal. Analyzing parallel programs with pin. *IEEE Computer*, 43(3):34–41, March 2010. DOI: 10.1109/MC.2010.60 Cited on page(s) 10

[20] D. F. Bacon, S. L. Graham, and O. J. Sharp. Compiler transformations for high-performance computing. *ACM Comput. Surv.*, 26:345–420, December 1994. DOI: 10.1145/197405.197406 Cited on page(s) 127

[21] U. Banerjee. Unimodular transformations of double loops. In *Third Workshop on Languages and Compilers for Parallel Computing*, pages 192–219. The MIT Press, 1990. Cited on page(s) 93, 94, 127

[22] U. Banerjee. *Dependence analysis (loop transformation for restructuring compilers*. Springer, 1996. Cited on page(s) 37, 125

[23] U. Banerjee, R. Eigenmann, A. Nicolau, and D. A. Padua. Automatic program parallelization. *Proceedings of the IEEE*, 81:211 – 243, February 1993. DOI: 10.1109/5.214548 Cited on page(s) 125, 126

[24] U. Banjerjee. *Dependence Analysis for Supercomputing*. Springer, 1988. Cited on page(s) 121, 125

[25] K. E. Batcher. Design of a massively parallel processor. *IEEE Transactions on Computers*, C29:836–840, September 1980. DOI: 10.1109/TC.1980.1675684 Cited on page(s) 7

[26] M.-W. Benabderrahmane, L.-N. Pouchet, A. Cohen, and C. Bastoul. The polyhedral model is more widely applicable than you think. In *Compiler Construction*, pages 283–303, 2010. DOI: 10.1007/978-3-642-11970-5_16 Cited on page(s) 126

[27] Beowulf.org. Cited on page(s) 9

[28] A. Bhowmik and M. Franklin. A general compiler framework for speculative multithreading. In *Proceedings of the fourteenth annual ACM Symposium on Parallel Algorithms and Architectures*, SPAA '02, pages 99–108, New York, NY, USA, 2002. ACM. DOI: 10.1145/564870.564885 Cited on page(s) 130

[29] IBM Research Blue Gene project page. `http://www.research.ibm.com/bluegene/press_release.html`. Last accessed January 5, 2012. Cited on page(s) 9, 95

[30] W. Blume and R. Eigenmann. Performance analysis of parallelizing compilers on the perfect benchmarks programs. *IEEE Transactions on Parallel and Distributed Systems*, 3:643–656, 1992. DOI: 10.1109/71.180621 Cited on page(s) 127

[31] W. Blume and R. Eigenmann. The Range Test: a dependence test for symbolic, non-linear expressions. In *Supercomputing '94*, pages 528–537, 1994. DOI: 10.1145/602770.602858 Cited on page(s) 63, 126

[32] H.-J. Boehm and S. V. Adve. Foundations of the C++ concurrency memory model. In *Proceedings of the ACM SIGPLAN 2008 Conference on Programming Language Design and Implementation*, pages 68–78, 2008. DOI: 10.1145/1375581.1375591 Cited on page(s) 126

[33] R. Bordawekar, U. Bondhugula, and R. Rao. Believe it or not!: multi-core cpus can match gpu performance for a flop-intensive application! In *International Conference on Parallel Architectures and Compilation Techniques*, pages 537–538, 2010. DOI: 10.1145/1854273.1854340 Cited on page(s) 126

[34] W. Bouknight, S. Denenberg, D. McIntyre, J. Randall, A. Sameh, and D. Slotnick. The Illiac IV system. *Proceedings of the IEEE*, 60(4):369–388, April 1972. DOI: 10.1109/PROC.1972.8647 Cited on page(s) 7

[35] Z. Bozkus, A. Choudhary, G. Fox, T. Haupt, and S. Ranka. Fortran 90D/HPF compiler for distributed memory MIMD computers: design, implementation, and performance results. In *Proceedings of the 1993 ACM/IEEE conference on Supercomputing*, Supercomputing '93, pages

351–360, New York, NY, USA, 1993. ACM. DOI: 10.1145/169627.169750 Cited on page(s) 110, 128

[36] Z. Bozkus, L. Meadows, S. Nakamoto, V. Schuster, and M. Young. PGHPF - an optimizing High Performance Fortran compiler for distributed memory machines. *Scientific Programming*, 6(1):29–40, 1997. Cited on page(s) 114, 128

[37] M. Bridges, N. Vachharajani, Y. Zhang, T. Jablin, and D. August. Revisiting the sequential programming model for multi-core. In *Proceedings of the 40th Annual IEEE/ACM International Symposium on Microarchitecture*, MICRO 40, pages 69–84, Washington, DC, USA, 2007. IEEE Computer Society. DOI: 10.1109/MICRO.2007.20 Cited on page(s) 131

[38] I. Buck, T. Foley, D. R. Horn, J. Sugerman, K. Fatahalian, M. Houston, and P. Hanrahan. Brook for gpus: stream computing on graphics hardware. *ACM Trans. Graph.*, 23(3):777–786, 2004. DOI: 10.1145/1015706.1015800 Cited on page(s) 7, 126

[39] M. Burke and R. Cytron. Interprocedural dependence analysis and parallelization. In *Proceedings of the ACM SIGPLAN'86 Symposium on Compiler Construction*, volume 21(6) of *SIGPLAN Notices*, page 162–175, June 1986. DOI: 10.1145/12276.13328 Cited on page(s) 33

[40] D. Callahan, J. Dongarra, and D. Levine. Vectorizing compilers: a test suite and results. In *Proceedings of the 1988 ACM/IEEE conference on Supercomputing*, Supercomputing '88, pages 98–105, Los Alamitos, CA, USA, 1988. IEEE Computer Society Press. DOI: 10.1109/SUPERC.1988.44642 Cited on page(s) 126

[41] D. Callahan and K. Kennedy. Estimating interlock and improving balance for pipelined machines. *Journal of Parallel and Distributed Computing*, 5:334–358, 1988. DOI: 10.1016/0743-7315(88)90002-0 Cited on page(s) 127

[42] D. Callahan, K. Kennedy, and J. Subhlok. Analysis of event synchronization in a parallel programming tool. In *PPOPP*, pages 21–30, 1990. DOI: 10.1145/99163.99167 Cited on page(s) 126

[43] W. W. Carlson, J. M. Draper, K. Y. D. E. Culler, E. Brooks, and K. Warren. Introduction to UPC and language specification. Technical report, IDA Center for Computing Sciences, 1999. Technical Report CCS-TR-99- 157. Cited on page(s) 128

[44] S. Carr, C. Ding, and P. Improving software pipelining with unroll-and-jam. In *28th Hawaii International Conference on System Sciences*, 1996. DOI: 10.1109/HICSS.1996.495462 Cited on page(s) 127

[45] S. Carr and K. Kennedy. Improving the ratio of memory operations to floating-point operations in loops. *ACM Transactions on Programming Languages and Systems*, 16(6):1768–1810, 1994. DOI: 10.1145/197320.197366 Cited on page(s) 127

[46] B. Chapman, P. Mehrotra, and H. Zima. Programming in Vienna Fortran. *Scientific Programming*, 1(1):31–50, 1992. Cited on page(s) 110, 128

[47] M. Chastain, G. Gostin, and J. M. S. Wallach. The Convex C240 architecture. In *Proceedings of the 1988 ACM/IEEE conference on Supercomputing*, Supercomputing '88, pages 321–329, Los Alamitos, CA, USA, 1988. IEEE Computer Society Press. DOI: 10.1109/SUPERC.1988.44669 Cited on page(s) 1

[48] S. Chatterjee, A. R. Lebeck, P. K. Patnala, and M. Thottethodi. Recursive array layouts and fast parallel matrix multiplication. In *Proceedings of Eleventh Annual ACM Symposium on Parallel Algorithms and Architectures*, pages 222–231, 1999. DOI: 10.1145/305619.305645 Cited on page(s) 128

[49] D.-K. Chen, H.-M. Su, and P.-C. Yew. The impact of synchronization and granularity on parallel systems. In *Proceedings of the 17th annual International Symposium on Computer Architecture*, ISCA '90, pages 239–248. ACM, 1990. DOI: 10.1109/ISCA.1990.134531 Cited on page(s) 126

[50] J.-H. Chow, L. E. Lyon, and V. Sarkar. Automatic parallelization for symmetric shared-memory multiprocessors. In *Proceedings of the 1996 conference of the Centre for Advanced Studies on Collaborative Research*, page 5. IBM, 1996. Cited on page(s) 4

[51] The Cilk project. http://supertech.csail.mit.edu/cilk/. Last accessed January 5, 2012. Cited on page(s) 63

[52] M. Cintra and D. R. Llanos. Toward efficient and robust software speculative parallelization on multiprocessors. In *Proceedings of the ninth ACM SIGPLAN Symposium on the Principles and Practice of Parallel Programming*, PPoPP '03, pages 13–24, New York, NY, USA, 2003. ACM. DOI: 10.1145/781498.781501 Cited on page(s) 129

[53] M. Cintra, J. F. Martínez, and J. Torrellas. Architectural support for scalable speculative parallelization in shared-memory multiprocessors. In *Proceedings of the 27th annual International Symposium on Computer Architecture*, pages 13–24, New York, NY, USA, 2000. ACM. DOI: 10.1145/342001.363382 Cited on page(s) 130

[54] Co-Array Fortran. http://www.co-array.org/. Last accessed August 26, 2011. Cited on page(s) 128

[55] C. Coarfa, Y. Dotsenko, J. Mellor-Crummey, F. Cantonnet, T. El-Ghazawi, A. Mohanti, Y. Yao, and D. Chavarría-Miranda. An evaluation of global address space languages: Co-array Fortran and Unified Parallel C. In *Proceedings of the tenth ACM SIGPLAN Symposium on the Principles and Practice of Parallel Programming*, PPoPP '05, pages 36–47, New York, NY, USA, 2005. ACM. DOI: 10.1145/1065944.1065950 Cited on page(s) 128

[56] J.-F. Collard. Automatic parallelization of while-loops using speculative execution. *International Journal of Parallel Programming*, 23:191–219, 1995. 10.1007/BF02577789. DOI: 10.1007/BF02577789 Cited on page(s) 127, 129

[57] K. Cooper and L. Torczon. *Engineering a Compiler*. Morgan Kaufmann, 2011. Cited on page(s) 125

[58] P. Cousot. Abstract interpretation. *ACM Comput. Surv.*, 28:324–328, June 1996. DOI: 10.1145/234528.234740 Cited on page(s) 28

[59] Abstract interpretation, 2008. http://www.di.ens.fr/~cousot/AI/. Last accessed January 5, 2012. Cited on page(s) 28

[60] CUDA: Parallel programming made easy. http://www.nvidia.com/object/cuda_home_new.html. Last accessed January 5, 2012. Cited on page(s) 7, 126

[61] R. Cytron. Doacross: Beyond vectorization for multiprocessors. In *International Conference on Parallel Processing*, pages 836–844, 1986. Cited on page(s) 58

[62] R. Cytron, J. Ferrante, B. K. Rosen, M. N. Wegman, and F. K. Zadeck. An efficient method of computing static single assignment form. In *ACM Conference on the Principals of Programming Languages*, pages 25–35, 1989. DOI: 10.1145/75277.75280 Cited on page(s) 13, 14, 38, 39, 125

[63] F. H. Dang and L. Rauchwerger. Speculative parallelization of partially parallel loops. In *Languages, Compilers, and Run-Time Systems for Scalable Computers*, volume 1915 of *Lecture Notes in Computer Science*, pages 285–299. Springer, 2000. Cited on page(s) 129

[64] F. H. Dang, H. Yu, and L. Rauchwerger. The R-LRPD test: Speculative parallelization of partially parallel loops. In *IPDPS*, 2002. DOI: 10.1109/IPDPS.2002.1015493 Cited on page(s) 129

[65] C. Ding, X. Shen, K. Kelsey, C. Tice, R. Huang, and C. Zhang. Software behavior oriented parallelization. In *Proceedings of the 2007 ACM SIGPLAN conference on Programming language design and implementation*, PLDI '07, pages 223–234, New York, NY, USA, 2007. ACM. DOI: 10.1145/1250734.1250760 Cited on page(s) 131

[66] Y. Dotsenko, C. Coarfa, and J. Mellor-Crummey. A multi-platform Co-Array Fortran compiler. In *Proceedings of the 13th International Conference on Parallel Architectures and Compilation Techniques*, PACT '04, pages 29–40, Washington, DC, USA, 2004. IEEE Computer Society. DOI: 10.1109/PACT.2004.1342539 Cited on page(s) 128

[67] Z.-H. Du, C.-C. Lim, X.-F. Li, C. Yang, Q. Zhao, and T.-F. Ngai. A cost-driven compilation framework for speculative parallelization of sequential programs. In *Proceedings of the ACM Conference on Programming Language Design and Implementation*, 2004. DOI: 10.1145/996841.996852 Cited on page(s) 130

[68] K. Ebcioglu. A compilation technique for software pipelining of loops with conditional jumps. In *20th Annual Workshop on microprogramming*, 1987. DOI: 10.1145/255305.255317 Cited on page(s) 127

[69] K. Ebcioglu, R. D. Groves, K.-C. Kim, G. M. Silberman, and I. Ziv. VLIW compilation techniques in a superscalar environment. In *Proceedings of the ACM Conference on Programming Language Design and Implementation*, pages 36–48, 1994. DOI: 10.1145/178243.178247 Cited on page(s) 67

[70] A. E. Eichenberger, P. Wu, and K. O'Brien. Vectorization for SIMD architectures with alignment constraints. In *Proceedings of the ACM SIGPLAN 2004 conference on Programming language design and implementation*, pages 82–93, New York, NY, USA, 2004. ACM. DOI: 10.1145/996841.996853 Cited on page(s) 126

[71] E. Elmroth and F. G. Gustavson. Applying recursion to serial and parallel qr factorization leads to better performance. *IBM Journal of Research and Development*, 44(4):605–624, 2000. DOI: 10.1147/rd.444.0605 Cited on page(s) 128

[72] P. Feautrier. Parametric integer programming. *RAIRO Recherche Op'erationnelle*, 22, 1988. Cited on page(s) 126

[73] P. Feautrier. Automatic parallelization in the polytope model. In G.-R. Perrin and A. Darte, editors, *The Data Parallel Programming Model*, volume 1132 of *Lecture Notes in Computer Science*, pages 79–103. Springer Berlin / Heidelberg, 1996. Cited on page(s) 126

[74] J. Ferrante, K. J. Ottenstein, and J. D. Warren. The program dependence graph and its use in optimization. In *6th International Symposium on Programming*, volume 167, pages 125–132, 1984. DOI: 10.1145/24039.24041 Cited on page(s) 125

[75] J. Ferrante, K. J. Ottenstein, and J. D. Warren. The program dependence graph and it use in optimization. *ACM Trans. Program. Lang. Syst.*, 9(3):319–349, 1987. DOI: 10.1145/24039.24041 Cited on page(s) 13, 38, 39

[76] FFTW. The project web page is at http://fftw.org/. Last accessed on January 5, 2012. Cited on page(s) 128

[77] C. N. Fischer, R. K. Cytron, and R. J. LeBlanc. *Crafting A Compiler*. Addison-Wesley Publishing Company, USA, 1st edition, 2009. Cited on page(s) 11, 19, 125

[78] A. Fisher and A. Ghuloum. Parallelizing complex scans and reductions. In *Conference on Programming Language Design and Implementation*, 1994. DOI: 10.1145/773473.178255 Cited on page(s) 127

[79] A. Fisher and A. Ghuloum. Parallelizing complex scans and reductions. *SIGPLAN Not.*, 29:135–146, June 1994. DOI: 10.1145/773473.178255 Cited on page(s) 127

[80] J. Fisher. Trace scheduling: a technique for global microcode compaction. *IEEE Trans. on Computers*, C-30(7), July 1981. DOI: 10.1109/TC.1981.1675827 Cited on page(s) 127

[81] J. A. Fisher, J. R. Ellis, J. C. Ruttenberg, and A. Nicolau. Parallel processing: a smart compiler and a dumb machine (with retrospective). In *20 Years of the ACM SIGPLAN Conference on Programming Language Design and Implementation 1979-1999, A Selection*, pages 112–124, 1984. DOI: 10.1145/989393.989408 Cited on page(s) 67, 127

[82] M. Frigo and S. G. Johnson. The design and implementation of FFTW3. *Proceedings of the IEEE*, 93(2):216–231, 2005. Special issue on "Program Generation, Optimization, and Platform Adaptation". DOI: 10.1109/JPROC.2004.840301 Cited on page(s) 128

[83] D. Gannon, W. Jalby, and K. Gallivan. Strategies for cache and local memory management by global program transformation. *J. Parallel Distrib. Comput.*, 5(5):587–616, 1988. DOI: 10.1016/0743-7315(88)90014-7 Cited on page(s) 127

[84] M. J. Garzarán, M. Prvulovic, V. Viñals, J. M. Llabería, L. Rauchwerger, and J. Torrellas. Using software logging to support multi-version buffering in thread-level speculation. In *IEEE PACT*, pages 170–179, 2003. DOI: 10.1109/PACT.2003.1238013 Cited on page(s) 129

[85] G. Goff, K. Kennedy, and C.-W. Tseng. Practical dependence testing. In *PLDI*, pages 15–29, 1991. DOI: 10.1145/113446.113448 Cited on page(s) 29

[86] M. Grigni and F. Manne. On the complexity of the generalized block distribution. In *Third International Workshop on Parallel Algorithms for Irregularly Structured Problems*, pages 319–326, 1996. Proceedings available in volume 1117 of Springer LNCS. DOI: 10.1007/BFb0030123 Cited on page(s) 128

[87] D. Grune, H. E. Bal, C. J. Jacobs, and K. G. Langendoen. *Modern Compiler Design*. Wiley, 2000. Cited on page(s) 11

[88] M. Gupta and P. Banerjee. Paradigm: a compiler for automatic data distribution on multi-computers. In *Proceedings of the 7th International Conference on Supercomputing*, ICS '93, pages 87–96, New York, NY, USA, 1993. ACM. DOI: 10.1145/165939.165959 Cited on page(s) 128

[89] M. Gupta, S. Midkiff, E. Schonberg, V. Seshadri, D. Shields, K.-Y. Wang, W.-M. Ching, and T. Ngo. An HPF compiler for the IBM SP2. In *Proceedings of the 1995 ACM/IEEE conference on Supercomputing (CDROM)*, Supercomputing '95, New York, NY, USA, 1995. ACM. DOI: 10.1145/224170.224422 Cited on page(s) 114, 128

[90] M. Gupta, S. Mukhopadhyay, and N. Sinha. Automatic parallelization of recursive procedures. In *Proceedings of the IEEE Conference on Parallel Architecture and Compiler Techniques (PACT)*, pages 139–148, 1999. DOI: 10.1023/A:1007560600904 Cited on page(s) 63

[91] M. Gupta, S. Mukhopadhyay, and N. Sinha. Automatic parallelization of recursive procedures. In *Proceedings of the 1999 International Conference on Parallel Architectures and Compilation Techniques*, PACT '99, pages 139–, Washington, DC, USA, 1999. IEEE Computer Society. DOI: 10.1109/PACT.1999.807504 Cited on page(s) 127

[92] M. Gupta, S. Mukhopadhyay, and N. Sinha. Automatic parallelization of recursive procedures. *International Journal of Parallel Programming*, 28(6):537–562, 2000. DOI: 10.1023/A:1007560600904 Cited on page(s) 63

[93] F. G. Gustavson, I. Jonsson, B. Kågström, and P. Ling. Towards peak performance on hierarchical smp memory architectures - new recursive blocked data formats and blas. In *PPSC*, 1999. Cited on page(s) 128

[94] M. Haghighat and C. Polychronopoulos. Symbolic program analysis and optimization for parallelizing compilers. In U. Banerjee, D. Gelernter, A. Nicolau, and D. Padua, editors, *Languages and Compilers for Parallel Computing*, volume 757 of *Lecture Notes in Computer Science*, pages 538–562. Springer Berlin / Heidelberg, 1993. Cited on page(s) 127

[95] M. Haghighat and C. Polychronopoulos. Symbolic analysis: A basis for parallelization, optimization, and scheduling of programs. In U. Banerjee, D. Gelernter, A. Nicolau, and D. Padua, editors, *Languages and Compilers for Parallel Computing*, volume 768 of *Lecture Notes in Computer Science*, pages 567–585. Springer Berlin / Heidelberg, 1994. Cited on page(s) 127

[96] M. R. Haghighat and C. D. Polychronopoulos. Symbolic program analysis and optimization for parallelizing compilers. In *Proceedings of the International Workshop on Languages and Compilers for Parallel Computers*, pages 538–562, 1992. Cited on page(s) 127

[97] R. A. Haring. Ibm blue gene/q compute chip. In *Hot Chips 23*, 2011. Available at http://www.hotchips.org/conference-archives/hot-chips-23. Last accessed January 5, 2012. Cited on page(s) 130

[98] J. Harris, J. A. Bircsak, M. R. Bolduc, J. A. Diewald, I. Gale, N. W. Johnson, S. Lee, C. A. Nelson, and C. D. Offner. Compiling High Performance Fortran for distributed-memory systems. *Digital Technical Journal*, 7(3):5–23, 1995. Cited on page(s) 114, 128

[99] T. Harris, J. R. Larus, and R. Rajwar. *Transactional Memory, 2nd edition*. Synthesis Lectures on Computer Architecture. Morgan & Claypool Publishers, 2010. Cited on page(s) 130

[100] M. S. Hecht. *Global data-flow analysis of computer programs*. PhD thesis, Princeton University, Princeton, NJ, USA, 1973. tech report no. AAI7409690. Cited on page(s) 19

[101] M. Herlihy and J. E. B. Moss. Transactional memory: Architectural support for lock-free data structures. In *Proceedings of the ACM International Symposium on Computer Architecture*, pages 289–300, 1993. DOI: 10.1109/ISCA.1993.698569 Cited on page(s) 66

[102] D. Hillis. *The Connection Machine*. MIT Press, 1989. Cited on page(s) 7

[103] S. Hiranandani, K. Kennedy, and C.-W. Tseng. Compiling Fortran D for MIMD distributed-memory machines. *Commun. ACM*, 35:66–80, August 1992. DOI: 10.1145/135226.135230 Cited on page(s) 110, 128

[104] J. Hoeflinger and Y. Paek. A comparative analysis of dependence testing mechanisms. In *Proceedings of the International Workshop on Languages and Compilers for Parallel Computers*, pages 289–303, 2000. DOI: 10.1007/3-540-45574-4_19 Cited on page(s) 126

[105] M. Hopkins. A perspective on the 801/Reduced Instruction Set Computer. *IBM Systems Journal*, 26(1):107–121, 1987. DOI: 10.1147/sj.261.0107 Cited on page(s) 58

[106] Power4 design. http://www.research.ibm.com/power4/. Last accessed January 5, 2012. Cited on page(s) 1

[107] E. B. III and K. Warren. The 1991 MPCI yearly report: The attack of the killer micros. Technical report, Lawrence Livermore National Laboratory, 1991. Technical Report UCRL-ID-107022. Cited on page(s) 1, 58

[108] W. L. H. III. The interprocedural analysis and automatic parallelization of Ccheme programs. *Lisp and Symbolic Computation*, 2(2):179–396, 1989. DOI: 10.1007/BF01808954 Cited on page(s) 62, 127

[109] W. L. H. III and Z. Ammarguellat. A program's eye view of Miprac. In *5th International Workshop on Languages and Compilers for Parallel Computing*, volume 757 of *Lecture Notes in Computer Science*. Springer, August 3-5 1993. Cited on page(s) 62, 127

[110] Intel Pentium D Processor 820. http://ark.intel.com/Product.aspx?id=27512. Last accessed January 5, 2012. Cited on page(s) 1

[111] Intel graphics media accelerator 950. http://www.intel.com/products/chipsets/gma950/index.htm. Last accessed January 5, 2012. Cited on page(s) 7

[112] G. Jin, J. Mellor-Crummey, and R. Fowler. Increasing temporal locality with skewing and recursive blocking. In *Proceedings of the 2001 ACM/IEEE conference on Supercomputing (CDROM)*, Supercomputing '01, pages 43–43, New York, NY, USA, 2001. ACM. DOI: 10.1145/582034.582077 Cited on page(s) 127

[113] R. B. Jones and V. H. Allan. Software pipelining: a comparison and improvement. In *Proceedings of the 23rd annual Workshop and Symposium on Microprogramming and Microarchitecture*, MICRO 23, pages 46–56, Los Alamitos, CA, USA, 1990. IEEE Computer Society Press. DOI: 10.1109/MICRO.1990.151426 Cited on page(s) 127

[114] P. Jouvelot and B. Dehbonei. A unified semantic approach for the vectorization and par-allelization of generalized reductions. In *1989 International Conference on Supercomputing*, 1989. DOI: 10.1145/318789.318810 Cited on page(s) 127

[115] A. Kejariwal, H. Saito, X. Tian, M. Girkar, W. Li, U. Banerjee, A. Nicolau, and C. n. D. Polychronopoulos. Lightweight lock-free synchronization methods for multithreading. In *Proceedings of the 20th annual International Conference on Supercomputing (ICS '06)*, ICS '06, pages 361–371, New York, NY, USA, 2006. ACM. DOI: 10.1145/1183401.1183452 Cited on page(s) 126

[116] K. Kennedy and R. Allen. *Optimizing Compilers for Modern Architectures: A Dependence-based Approach*. Morgan Kaufmann, 2001. Cited on page(s) 126

[117] K. Kennedy, C. Koelbel, and H. Zima. The rise and fall of High Performance Fortran: an historical object lesson. In *Proceedings of the third ACM SIGPLAN conference on History of programming languages*, HOPL III, pages 712–722, New York, NY, USA, 2007. ACM. DOI: 10.1145/1238844.1238851 Cited on page(s) 128

[118] K. Kennedy and U. Kremer. Automatic data layout for distributed-memory machines. *ACM Trans. Program. Lang. Syst.*, 20:869–916, July 1998. DOI: 10.1145/291891.291901 Cited on page(s) 128

[119] D. Klappholz, K. Psarris, and X. Kong. On the perfect accuracy of an approximate sub-script analysis test. In *International Conference on Supercomputing*, pages 201–212, 1990. DOI: 10.1145/255129.255158 Cited on page(s) 37

[120] K. Knobe and V. Sarkar. Array SSA form and its use in parallelization. In *ACM Conference on the Principals of Programming Languages*, pages 107–120, 1998. DOI: 10.1145/268946.268956 Cited on page(s) 13

[121] I. Kodukula, N. Ahmed, and K. Pingali. Data-centric multi-level blocking. In *Proceedings of the ACM Conference on Programming Language Design and Implementation*, pages 346–357, 1997. DOI: 10.1145/258915.258946 Cited on page(s) 127

[122] C. Koelbel. Compile-time generation of regular communications patterns. In *Proceedings of the 1991 ACM/IEEE conference on Supercomputing*, Supercomputing '91, pages 101–110, New York, NY, USA, 1991. ACM. DOI: 10.1145/125826.125890 Cited on page(s) 128

[123] C. Koelbel. An overview of High Performance Fortran. *SIGPLAN Fortran Forum*, 11:9–16, December 1992. DOI: 10.1145/140734.140736 Cited on page(s) 128

[124] C. Koelbel, D. Loveman, R. Schreiber, G. Steele, and M. E. Zosel. *The High Performance Fortran Handbook*. MIT Press, 1993. Cited on page(s) 101, 128

[125] X. Kong, D. Klappholz, and K. Psarris. The I test: an improved dependence test for automatic parallelization and vectorization. *Parallel and Distributed Systems, IEEE Transactions on*, 2(3):342 –349, jul 1991. DOI: 10.1109/71.86109 Cited on page(s) 126

[126] J. S. Kowalik. *Parallel MIMD computation : The HEP supercomputer and its applications*. MIT Press, 1985. available as `http://hdl.handle.net/1721.1/1745`. Last accessed January 5, 2012. Cited on page(s) 56

[127] A. Krishnamurthy and K. A. Yelick. Optimizing parallel programs with explicit synchronization. In *Proceedings of the ACM SIGPLAN Conference on Programming Language Design and Implementation*, pages 196–204, 1995. DOI: 10.1145/223428.207142 Cited on page(s) 126

[128] A. Krishnamurthy and K. A. Yelick. Analyses and optimizations for shared address space programs. *J. Parallel Distrib. Comput.*, 38(2):130–144, 1996. DOI: 10.1006/jpdc.1996.0136 Cited on page(s) 46, 50, 126

[129] D. Kuck and A. Goldstein. `http://www.ieeeghn.org/wiki/index.php/Oral-History:David_Kuck`. Last accessed January 5, 2012. Cited on page(s) 1

[130] D. J. Kuck. *The Structure of Computers and Computations*. John Wiley & Sons, Inc., 1978. Cited on page(s) 127

[131] D. J. Kuck, E. S. Davidson, D. H. Lawrie, A. H. Sameh, and C.-Q. Zhu. The Cedar system and an initial performance study. In *25 Years of ISCA: Retrospectives and Reprints*, pages 462–472, 1998. DOI: 10.1145/285930.286005 Cited on page(s) 126

[132] D. J. Kuck, R. H. Kuhn, D. A. Padua, B. Leasure, and M. Wolfe. Dependence graphs and compiler optimizations. In *ACM Conference on the Principles of Programming Languages*, pages 207–218, 1981. DOI: 10.1145/567532.567555 Cited on page(s) 1, 126

[133] D. J. Kuck, R. H. Kuhn, D. A. Padua, B. Leasure, and M. Wolfe. Dependence graphs and compiler optimizations. In *Proceedings of the ACM Conference on Principles of Programming Languages*, pages 207–218, 1981. DOI: 10.1145/567532.567555 Cited on page(s) 127

[134] R. H. Kuhn. *Optimization and interconnection complexity for: Parallel processors, single stage networks, and decision trees*. PhD thesis, Department of Computer Science, University of Illinois at Urbana-Champaign, February 1980. Cited on page(s) 37

[135] M. Kulkarni, K. Pingali, G. Ramanarayanan, B. Walter, K. Bala, and L. P. Chew. Optimistic parallelism benefits from data partitioning. In *Proceedings of the 13th International Conference on Architectural Support for Programming Languages and Operating Systems (ASPLOS '08)*, ASPLOS XIII, pages 233–243, New York, NY, USA, 2008. ACM. DOI: 10.1145/1346281.1346311 Cited on page(s) 131

[136] M. Kulkarni, K. Pingali, B. Walter, G. Ramanarayanan, K. Bala, and L. P. Chew. Optimistic parallelism requires abstractions. In *Proceedings of the 2007 ACM SIGPLAN conference on Programming language design and implementation*, PLDI '07, pages 211–222, New York, NY, USA, 2007. ACM. DOI: 10.1145/1250734.1250759 Cited on page(s) 66, 131

[137] M. Kulkarni, K. Pingali, B. Walter, G. Ramanarayanan, K. Bala, and L. P. Chew. Optimistic parallelism requires abstractions. *Communications of the ACM*, 52:89–97, Sept. 2009. DOI: 10.1145/1562164.1562188 Cited on page(s) 66, 131

[138] M. S. Lam. Software pipelining: an effective scheduling technique for VLIW machines (with retrospective). In *20 Years of the ACM SIGPLAN Conference on Programming Language Design and Implementation 1979-1999, A Selection*, pages 244–256. ACM, 2004. DOI: 10.1145/989393.989420 Cited on page(s) 127

[139] M. S. Lam and M. E. Wolf. A data locality optimizing algorithm (with retrospective). In *20 Years of the ACM SIGPLAN Conference on Programming Language Design and Implementation 1979-1999, A Selection*, pages 442–459, 1991. DOI: 10.1145/989393.989437 Cited on page(s) 29

[140] M. S. Lam and M. E. Wolf. A data locality optimizing algorithm. *SIGPLAN Not.*, 39:442–459, April 2004. DOI: 10.1145/989393.989437 Cited on page(s) 127

[141] J. Lee, D. A. Padua, and S. P. Midkiff. Basic compiler algorithms for parallel programs. In *Proceedings of the seventh ACM SIGPLAN Symposium on the Principles and Practice of Parallel Programming*, PPoPP '99, pages 1–12, 1999. DOI: 10.1145/301104.301105 Cited on page(s) 11, 126

[142] S. Lee, S.-J. Min, and R. Eigenmann. Openmp to gpgpu: a compiler framework for automatic translation and optimization. In *Proceedings of the 14th ACM SIGPLAN Symposium on Principles and Practice of Parallel Programming (PPOPP '09)*, pages 101–110, 2009. DOI: 10.1145/1594835.1504194 Cited on page(s) 7, 126

[143] The lex & yacc page. http://dinosaur.compilertools.net/. Last accessed January 5, 2012. Cited on page(s) 11

[144] Z. Li and W. A. Abu-Sufah. A technique for reducing synchronization overhead in large scale multiprocessors. In *Proceedings of the International Symposia on Computer Architecture*, pages 284–291, 1985. DOI: 10.1145/327070.327266 Cited on page(s) 62, 126

[145] H. Lin, S. P. Midkiff, and R. Eigenmann. A study of the usefulness of producer/consumer synchronization. In *Proceedings of the 24th International Workshop on Languages and Compilers for Parallel Computing*, 2011. To appear. Cited on page(s) 60

[146] W. Liu, J. Tuck, L. Ceze, W. Ahn, K. Strauss, J. Renau, and J. Torrellas. Posh: a tls compiler that exploits program structure. In *Proceedings of the eleventh ACM SIGPLAN Symposium on the Principles and Practice of Parallel Programming*, PPoPP '06, pages 158–167, New York, NY, USA, 2006. ACM. DOI: 10.1145/1122971.1122997 Cited on page(s) 130

[147] J. Manson, W. Pugh, and S. V. Adve. The Java memory model. *Proceedings of the 32nd ACM SIGPLAN-SIGACT Symposium on the Principles of Programming Languages*, 40:378–391, January 2005. Cited on page(s) 126

[148] P. Marcuello and A. Gonzalez. Thread-spawning schemes for speculative multithreading. In *High-Performance Computer Architecture, 2002. Proceedings. Eighth International Symposium on*, pages 55 – 64, feb. 2002. DOI: 10.1109/HPCA.2002.995698 Cited on page(s) 130

[149] W. R. Mark, R. S. Glanville, K. Akeley, and M. J. Kilgard. Cg: a system for programming graphics hardware in a c-like language. *ACM Trans. Graph.*, 22(3):896–907, 2003. DOI: 10.1145/882262.882362 Cited on page(s) 7, 126

[150] K. S. McKinley, S. Carr, and C.-W. Tseng. Improving data locality with loop transformations. *ACM Trans. Program. Lang. Syst.*, 18:424–453, July 1996. DOI: 10.1145/233561.233564 Cited on page(s) 127

[151] M. Mehrara, J. Hao, P.-C. Hsu, and S. Mahlke. Parallelizing sequential applications on commodity hardware using a low-cost software transactional memory. In *Proceedings of the 2009 ACM SIGPLAN conference on Programming language design and implementation*, PLDI '09, pages 166–176, New York, NY, USA, 2009. ACM. DOI: 10.1145/1542476.1542495 Cited on page(s) 130

[152] S. P. Midkiff. Dependence analysis in parallel loops with i±k subscripts. In *Proceedings of the Eight International Workshop on Languages and Compilers for Parallel Computing*, volume 1033 of *Lecture Notes in Computer Science*, pages 331–345. Springer, 1995. DOI: 10.1007/BFb0014209 Cited on page(s) 46

[153] S. P. Midkiff. Local iteration set computation for block-cyclic distributions. In *Proceedings of the International Conference on Parallel Processing, Volume 2*, pages 77–84, 1995. Cited on page(s) 101

[154] S. P. Midkiff. Optimizing the representation of local iteration sets and access sequences for block-cyclic distributions. In *Ninth International Workshop on Languages and Compilers for Parallel Computing*, pages 420–434, 1996. DOI: 10.1007/BFb0017267 Cited on page(s) 101

[155] S. P. Midkiff, J. E. Moreira, and M. Snir. Optimizing array reference checking in Java programs. *IBM Systems Journal*, 37(3):409–453, 1998. DOI: 10.1147/sj.373.0409 Cited on page(s) 38

[156] S. P. Midkiff and D. A. Padua. Compiler generated synchronization for do loops. In *Proceedings of the International Conference on Parallel Programming (ICPP)*, pages 544–551, 1986. Cited on page(s) 61, 62, 126

[157] S. P. Midkiff and D. A. Padua. Compiler algorithms for synchronization. *IEEE Trans. Computers*, 36(12):1485–1495, 1987. DOI: 10.1109/TC.1987.5009499 Cited on page(s) 62, 126

[158] M. J. Moravan, J. Bobba, K. E. Moore, L. Yen, M. D. Hill, B. Liblit, M. M. Swift, and D. A. Wood. Supporting nested transactional memory in logtm. In *Proceedings of the 12th International Conference on Architectural Support for Programming Languages and Operating Systems, ASPLOS 2006, San Jose, CA, USA, October 21-25, 2006*, pages 359–370, 2006. DOI: 10.1145/1168918.1168902 Cited on page(s) 66

[159] J. E. Moreira, S. P. Midkiff, and M. Gupta. From flop to megaflops: Java for technical computing. *ACM Trans. Program. Lang. Syst.*, 22(2):265–295, 2000. DOI: 10.1145/349214.349222 Cited on page(s) 38

[160] Message passing interface forum. http://mpi-forum.org/. Last accessed January 5, 2012. Cited on page(s) 128

[161] S. S. Muchnick. *Advanced Compiler Design & Implementation*. Morgan Kaufmann/ Elsevier Science India, 2003. Cited on page(s) 125

[162] e. Narasimha R. Adiga. An overview of the BlueGene/L supercomputer. In *ACM Supercomputing (SC '02)*, pages 1–22, 2002. DOI: 10.1109/SC.2002.10017 Cited on page(s) 9, 95

[163] R. W. Numrich and J. Reid. Co-array Fortran for parallel programming. *SIGPLAN Fortran Forum*, 17:1–31, August 1998. DOI: 10.1145/289918.289920 Cited on page(s) 128

[164] Nvidia technologies. http://www.nvidia.com/page/technologies.html. Last accessed January 5, 2012. Cited on page(s) 7

[165] J. T. Oplinger, D. L. Heine, and M. S. Lam. In search of speculative thread-level parallelism. In *Proceedings of the 1999 International Conference on Parallel Architectures and Compilation Techniques*, PACT '99, pages 303–, Washington, DC, USA, 1999. IEEE Computer Society. DOI: 10.1109/PACT.1999.807576 Cited on page(s) 130

[166] G. Ottoni, R. Rangan, A. Stoler, and D. I. August. Automatic thread extraction with decoupled software pipelining. In *Proceedings of the 38th annual IEEE/ACM International Symposium on Microarchitecture*, MICRO 38, pages 105–118, Washington, DC, USA, 2005. IEEE Computer Society. DOI: 10.1109/MICRO.2005.13 Cited on page(s) 131

[167] D. Padua. The Fortran I compiler. *Computing in Science and Engg.*, 2:70–75, January 2000. DOI: 10.1109/5992.814661 Cited on page(s) 125

[168] D. A. Padua. *Multiprocessors: Discussion of Some Theoretical and Practical Problems*. PhD thesis, Dept. of Computer Science, Univ. of Illinois at Urbana-Champaign, October 1979. Cited on page(s) 58

[169] D. A. Padua and M. J. Wolfe. Advanced compiler optimizations for supercomputers. *Commun. ACM*, 29:1184–1201, December 1986. DOI: 10.1145/7902.7904 Cited on page(s) 125, 126, 127

[170] D. Palermo, E. Su, I. E.W. Hodges, and P. Banerjee. (R) compiler support for privatization on distributed - memory machines. *Parallel Processing, International Conference on*, 3:0017, 1996. DOI: 10.1109/ICPP.1996.538555 Cited on page(s) 127

[171] J. Patel and E. Davidson. Improving the throughput of a pipeline by insertion of delays. In *Third Annual Symposium on Computer Architecture*, pages 159–164, 1976. DOI: 10.1145/800110.803575 Cited on page(s) 127

[172] D. A. Patterson and C. H. Sequin. RISC I: a reduced instruction set VLSI computer. In *25 years of the International Symposia on Computer Architecture (selected papers)*, pages 216–230, New York, NY, USA, 1998. ACM. Originally appeared in ISCA '98. DOI: 10.1145/285930.285981 Cited on page(s) 58

[173] G. Pfister. *In Search of Clusters, Second Edition*. Prentice Hall, 1997. Cited on page(s) 9, 95

[174] Pin: A dynamic binary instrumentation tool. http://www.pintool.org/. Last accessed January 5, 2012. Cited on page(s) 10

[175] W. Pottenger. Induction variable substitution and reduction recognition in the polaris parallelizing compiler. MS Thesis. Cited on page(s) 127

[176] W. M. Pottenger and R. Eigenmann. Idiom recognition in the Polaris parallelizing compiler. In *International Conference on Supercomputing*, pages 444–448, 1995. DOI: 10.1145/224538.224655 Cited on page(s) 127

[177] L.-N. Pouchet, C. Bastoul, A. Cohen, and J. Cavazos. Iterative optimization in the polyhedral model: Part ii, multidimensional time. In *Proceedings of the ACM Conference on Programming Language Design and Implementation*, pages 90–100, 2008. DOI: 10.1145/1379022.1375594 Cited on page(s) 126

[178] L.-N. Pouchet, C. Bastoul, A. Cohen, and N. Vasilache. Iterative optimization in the polyhedral model: Part i, one-dimensional time. In *CGO*, pages 144–156, 2007. DOI: 10.1109/CGO.2007.21 Cited on page(s) 126

[179] Power.org. `http://www-03.ibm.com/systems/power/`. Last accessed January 5, 2012. Cited on page(s) 9

[180] IBM Power 770 and 780 (9117-MMC, 9179-MHC) technical overview and introduction. `http://www.redbooks.ibm.com/Redbooks.nsf/RedbookAbstracts/redp4798.html?Open`. Last accessed January 5, 2012. Cited on page(s) 9

[181] W. Pugh. The Omega Test: a fast and practical integer programming algorithm for dependence analysis. In *SC*, pages 4–13, 1991. DOI: 10.1145/125826.125848 Cited on page(s) 37, 126

[182] W. Pugh. The omega test: a fast and practical integer programming algorithm for dependence analysis. *Communications of the ACM*, (8):102 – 114, August 1992. DOI: 10.1145/125826.125848 Cited on page(s) 37

[183] W. Pugh and D. Wonnacott. Eliminating false data dependences using the Omega Test. In *Proceedings of the ACM Conference on Programming Language Design and Implementation*, pages 140–151, 1992. DOI: 10.1145/143103.143129 Cited on page(s) 126

[184] W. Pugh and D. Wonnacott. Going beyond integer programming with the Omega Test to eliminate false data dependences. *IEEE Trans. Parallel Distrib. Syst.*, 6(2):204–211, 1995. DOI: 10.1109/71.342135 Cited on page(s) 37, 126

[185] The Java memory model. A detailed discussion of problems and issues with the original Java memory model can be found at `http://www.cs.umd.edu/~pugh/java/memoryModel/`. Last accessed on January 5, 2012. Cited on page(s) 126

[186] C. G. Quiñones, C. Madriles, J. Sánchez, P. Marcuello, A. González, and D. M. Tullsen. Mitosis compiler: an infrastructure for speculative threading based on pre-computation slices. In *Proceedings of the 2005 ACM SIGPLAN conference on Programming language design and implementation*, PLDI '05, pages 269–279, New York, NY, USA, 2005. ACM. DOI: 10.1145/1064978.1065043 Cited on page(s) 130

[187] M. Quinn. *Parallel Programming in C with MPI and OpenMP*. McGraw-Hill Science/Engineering/Math, 2003. Cited on page(s) 100

[188] A. Raman, H. Kim, T. R. Mason, T. B. Jablin, and D. I. August. Speculative parallelization using software multi-threaded transactions. In *Proceedings of the fifteenth edition of ASPLOS on Architectural support for programming languages and operating systems*, ASPLOS '10, pages 65–76, New York, NY, USA, 2010. ACM. DOI: 10.1145/1736020.1736030 Cited on page(s) 131

[189] E. Raman, G. Ottoni, A. Raman, M. J. Bridges, and D. I. August. Parallel-stage decoupled software pipelining. In *Proceedings of the 6th annual IEEE/ACM International Symposium*

on Code generation and optimization, CGO '08, pages 114–123, New York, NY, USA, 2008. ACM. DOI: 10.1145/1356058.1356074 Cited on page(s) 55, 131

[190] R. Rangan, N. Vachharajani, M. Vachharajani, and D. I. August. Decoupled software pipelining with the synchronization array. In *Proceedings of the 13th International Conference on Parallel Architectures and Compilation Techniques (PACT 2004)*, pages 177–188, 2004. DOI: 10.1109/PACT.2004.1342552 Cited on page(s) 55, 131

[191] B. Rau and C. Glaeser. Some scheduling techniques and and easily horizontal architecture for high performance scientific computing. In *1th Annual Workshop on Microprogramming*, 1981. Cited on page(s) 127

[192] L. Rauchwerger, F. Arzu, and K. Ouchi. Standard templates adaptive parallel library (stapl). In *4th International Workshop on Languages, Compilers, and Run-Time Systems for Scalable Computers*, pages 402–409, 1998. Cited on page(s) 125

[193] L. Rauchwerger and D. A. Padua. Parallelizing while loops for multiprocessor systems. In *Proceedings of the 9th International Parallel Processing Symposium, April 25-28, 1995, Santa Barbara, California, USA (IPPS)*, pages 347–356, 1995. DOI: 10.1109/IPPS.1995.395955 Cited on page(s) 65, 127, 129

[194] L. Rauchwerger and D. A. Padua. The lrpd test: Speculative run-time parallelization of loops with privatization and reduction parallelization. *IEEE Trans. Parallel Distrib. Syst.*, 10(2):160–180, 1999. DOI: 10.1109/71.752782 Cited on page(s) 129

[195] J. Reid. Coarrays in the next Fortran standard. *SIGPLAN Fortran Forum*, 29:10–27, July 2010. DOI: 10.1145/1837137.1837138 Cited on page(s) 114, 128

[196] G. Ren, P. Wu, and D. Padua. A preliminary study on the vectorization of multimedia applications for multimedia extensions. In L. Rauchwerger, editor, *Languages and Compilers for Parallel Computing*, volume 2958 of *Lecture Notes in Computer Science*, pages 420–435. Springer Berlin / Heidelberg, 2004. DOI: 10.1007/978-3-540-24644-2_27 Cited on page(s) 126

[197] M. C. Rinard and P. C. Diniz. Commutativity analysis: A new analysis technique for parallelizing compilers. *ACM Trans. Program. Lang. Syst.*, 19(6):942–991, 1997. DOI: 10.1145/267959.269969 Cited on page(s) 63

[198] A. Rogers and K. Pingali. Process decomposition through locality of reference. In *Proceedings of the 1989 ACM Conference on Programming Language Design and Implementation*, pages 69–80, 1989. DOI: 10.1145/73141.74824 Cited on page(s) 97, 107

[199] H. Rong, Z. Tang, R. Govindarajan, A. Douillet, and G. R. Gao. Single-dimension software pipelining for multi-dimensional loops. In *Symposium on Code Generation and Optimization*, pages 163–174, 2004. DOI: 10.1145/1216544.1216550 Cited on page(s) 127

[200] B. K. Rosen. High-level data flow analysis. *Commun. ACM*, 20:712–724, October 1977. DOI: 10.1145/359842.359849 Cited on page(s) 19

[201] R. Rugina and M. Rinard. Automatic parallelization of divide and conquer algorithms. In *Proceedings of the seventh ACM SIGPLAN Symposium on the Principles and Practice of Parallel Programming*, PPoPP '99, pages 72–83, New York, NY, USA, 1999. ACM. DOI: 10.1145/301104.301111 Cited on page(s) 127

[202] S. Rus, L. Rauchwerger, and J. Hoeflinger. Hybrid analysis: static & dynamic memory reference analysis. In *ICS*, pages 274–284, 2002. DOI: 10.1023/A:1024597010150 Cited on page(s) 37, 126

[203] S. Rus, L. Rauchwerger, and J. Hoeflinger. Hybrid analysis: Static & dynamic memory reference analysis. *International Journal of Parallel Programming*, 31(4):251–283, 2003. Cited on page(s) 126

[204] R. Russell. The CRAY-1 computer system. *Communications of the ACM*, 21(1):63–72, January 1978. DOI: 10.1145/359327.359336 Cited on page(s) 1

[205] S. Ryoo, C. I. Rodrigues, S. S. Baghsorkhi, S. S. Stone, D. B. Kirk, and W. mei W. Hwu. Optimization principles and application performance evaluation of a multithreaded gpu using cuda. In *Proceedings of the 13th ACM SIGPLAN Symposium on Principles and Practice of Parallel Programming (PPOPP '08)*, pages 73–82, 2008. DOI: 10.1145/1345206.1345220 Cited on page(s) 7, 126

[206] J. Sampson, R. Gonzalez, J.-F. Collard, N. P. Jouppi, M. Schlansker, and B. Calder. Exploiting fine-grained data parallelism with chip multiprocessors and fast barriers. In *Proceedings of the 39th Annual IEEE/ACM International Symposium on Microarchitecture*, MICRO 39, pages 235–246, Washington, DC, USA, 2006. IEEE Computer Society. DOI: 10.1109/MICRO.2006.23 Cited on page(s) 126

[207] Intel Core i5-2540M processor. `http://ark.intel.com/products/50072/Intel-Core-i5-2540M-Processor-(3M-Cache-2_60-GHz)`. Cited on page(s) 9

[208] D. A. Schmidt. Data flow analysis is model checking of abstract interpretations. In *1998 ACM Conference on the Principles of Programming Languages (POPL '98)*, pages 38–48, 1998. DOI: 10.1145/268946.268950 Cited on page(s) 29

[209] D. Shasha and M. Snir. Efficient and correct execution of parallel programs that share memory. *ACM Trans. Program. Lang. Syst.*, 10(2):282–312, 1988. DOI: 10.1145/42190.42277 Cited on page(s) 43, 44, 45, 46, 126

[210] Z. Shen, Z. Li, and P. Yew. An empirical study of Fortran programs for parallelizing compilers. *IEEE Transactions on Parallel and Distributed Systems*, 1:356–364, 1990. DOI: 10.1109/71.80162 Cited on page(s) 127

[211] B. J. Smith. A pipelined, shared resource computer. In *Proceedings of the International Conference on Parallel Processing*, pages 6–8, 1978. Cited on page(s) 1

[212] G. Snelting, T. Robschink, and J. Krinke. Efficient path conditions in dependence graphs for software safety analysis. *ACM Trans. Softw. Eng. Methodol.*, 15:410–457, October 2006. DOI: 10.1145/1178625.1178628 Cited on page(s) 34

[213] M. Snir, J. Dongarra, J. S. Kowalik, S. Huss-Lederman, S. W. Otto, and D. W. Walker. *MPI: The Complete Reference*. MIT Press, 2nd edition, 1998. Cited on page(s) 128

[214] N. Sreraman and R. Govindarajan. A vectorizing compiler for multimedia extensions. *International Journal of Parallel Programming*, 28:363–400, 2000. 10.1023/A:1007559022013. DOI: 10.1023/A:1007559022013 Cited on page(s) 126

[215] SSE performance programming. http://developer.apple.com/hardwaredrivers/ve/sse.html. Last accessed January 5, 2012. Cited on page(s) 7

[216] Intel compiler options for intel SSE and intel AVX generation (SSE2, SSE3, SSE3_ATOM, SSSE3, SSE4.1, SSE4.2, AVX, AVX2) and processor-specific optimizations. http://software.intel.com/en-us/articles/performance-tools-for-software-developers-intel-compiler-options-for-sse-generation-and-processor-specific-optimizations/. Last accessed January 5, 2012. Cited on page(s) 7

[217] STAPL: Standard Template Adaptive Parallel Library. https://parasol.tamu.edu/groups/rwergergroup/research/stapl/. Last accessed January 5, 2012. DOI: 10.1145/1815695.1815713 Cited on page(s) 125

[218] J. Steffan and T. Mowry. The potential for using thread-level data speculation to facilitate automatic parallelization. In *Proceedings of the Symposium on High-Performance Computer Architecture*, pages 2 –13, feb 1998. DOI: 10.1109/HPCA.1998.650541 Cited on page(s) 129

[219] J. G. Steffan, C. Colohan, A. Zhai, and T. C. Mowry. The stampede approach to thread-level speculation. *ACM Trans. Comput. Syst.*, 23:253–300, August 2005. DOI: 10.1145/1082469.1082471 Cited on page(s) 131

[220] J. G. Steffan, C. B. Colohan, A. Zhai, and T. C. Mowry. Improving value communication for thread-level speculation. In *In Proceedings of the 8th International Symposium on High Performance Computer Architecture (HPCA)*, page 65, 2002. DOI: 10.1109/HPCA.2002.995699 Cited on page(s) 131

[221] B. Steffen. Data flow analysis as model checking. In *Proceedings of 1991 Conference on the Theoretical Aspects of Computer Science (TACS '91)*. Springer-Verlag, 1991. Available as volume 526 of Lecture Notes in Computer Science. DOI: 10.1007/3-540-54415-1_54 Cited on page(s) 29

[222] B. Steffen. Generating data-flow analysis algorithms for modal specifications. *Science of Computer Programming*, (139):21–115, 1993. DOI: 10.1016/0167-6423(93)90003-8 Cited on page(s) 29

[223] H.-M. Su and P.-C. Yew. On data synchronization for multiprocessors. In *International Symposium on Computer Architecture*, volume 17, pages 416–423, 1989. DOI: 10.1145/74926.74972 Cited on page(s) 126

[224] J. Su and K. A. Yelick. Automatic communication performance debugging in PGAS languages. In *Languages and Compilers for Parallel Computing*, pages 232–245, 2007. DOI: 10.1007/978-3-540-85261-2_16 Cited on page(s) 126

[225] Z. Sura, X. Fang, C.-L. Wong, S. P. Midkiff, J. Lee, and D. A. Padua. Compiler techniques for high performance sequentially consistent Java programs. In *Proceedings of the ACM SIGPLAN Symposium on Principles and Practice of Parallel Programming (PPoPP 2005)*, pages 2–13, 2005. DOI: 10.1145/1065944.1065947 Cited on page(s) 49, 126

[226] G. Tanase, A. A. Buss, A. Fidel, Harshvardhan, I. Papadopoulos, O. Pearce, T. G. Smith, N. Thomas, X. Xu, N. Mourad, J. Vu, M. Bianco, N. M. Amato, and L. Rauchwerger. The STAPL parallel container framework. In *Proceedings of the 16th ACM SIGPLAN Symposium on Principles and Practice of Parallel Programming*, pages 235–246, 2011. DOI: 10.1145/2038037.1941586 Cited on page(s) 125

[227] D. Tarditi, S. Puri, and J. Oglesby. Accelerator: using data parallelism to program gpus for general-purpose uses. In *Proceedings of the 12th International Conference on Architectural support for Programming Languages and Operating Systems*, ASPLOS-XII, pages 325–335, New York, NY, USA, 2006. ACM. DOI: 10.1145/1168919.1168898 Cited on page(s) 7, 126

[228] J. Test, M. Myszewski, and R. Swift. The alliant fx/series: A language driven architecture for parallel processing of dusty deck fortran. In J. de Bakker, A. Nijman, and P. Treleaven, editors, *PARLE Parallel Architectures and Languages Europe*, volume 258 of *Lecture Notes in Computer Science*, pages 345–356. Springer Berlin / Heidelberg, 1987. 10.1007/3-540-17943-7 Cited on page(s) 1, 58

[229] Intel threading building blocks 3.0 for open source. http://threadingbuildingblocks.org/. Last accessed January 5, 2012. Cited on page(s) 125

[230] X. Tian, A. Bik, M. Girkar, P. Grey, H. Saito, and E. Su. Intel OpenMP C++/Fortran compiler for hyper-threading technology: Implementation and performance. *Intel Technology Journal*, 6(01):1–11, 2002. Available at http://download.intel.com/technology/itj/2002/volume06issue01/art04_fortrancompiler/vol6iss1_art04.pdf, last checked January 5, 2012. Cited on page(s) 4

[231] G. Tournavitis, Z. Wang, B. Franke, and M. F. O'Boyle. Towards a holistic approach to auto-parallelization: integrating profile-driven parallelism detection and machine-learning based mapping. In *Proceedings of the 2009 ACM SIGPLAN conference on Programming language design and implementation*, PLDI '09, pages 177–187, New York, NY, USA, 2009. ACM. DOI: 10.1145/1542476.1542496 Cited on page(s) 130

[232] R. Touzeau. A Fortran compiler for the FPS-164 Scientific Computer. In *SIGPLAN 1984 Symposium on Compiler Construction*, pages 48 – 57, 1984. DOI: 10.1145/502874.502879 Cited on page(s) 127

[233] P. Tu and D. A. Padua. Automatic array privatization. In *Proceedings of the International Workshop on Languages and Compilers for Parallel Computers*, pages 500–521, 1993. DOI: 10.1007/3-540-45403-9_8 Cited on page(s) 127

[234] P. Tu and D. A. Padua. Gated SSA-based demand-driven symbolic analysis for parallelizing compilers. In *International Conference on Supercomputing*, pages 414–423, 1995. DOI: 10.1145/224538.224648 Cited on page(s) 127

[235] M. Ujaldon, E. L. Zapata, B. M. Chapman, and H. P. Zima. Vienna-fortran/hpf extensions for sparse and irregular problems and their compilation. *IEEE Trans. Parallel Distrib. Syst.*, 8(10):1068–1083, 1997. DOI: 10.1109/71.629489 Cited on page(s) 128

[236] G. Upadhyaya, S. P. Midkiff, and V. S. Pai. Using data structure knowledge for efficient lock generation and strong atomicity. In *Proceedings of the 15th ACM SIGPLAN Symposium on Principles and Practice of Parallel Programming*, PPOPP 2010, pages 281–292, 2010. DOI: 10.1145/1837853.1693490 Cited on page(s) 66

[237] Berkeley unified parallel C. http://upc.lbl.gov/. Last accessed January 5, 2012. Cited on page(s) 50, 128

[238] N. Vachharajani, R. Rangan, E. Raman, M. J. Bridges, G. Ottoni, and D. I. August. Speculative decoupled software pipelining. In *Proceedings of the 16th International Conference on Parallel Architecture and Compilation Techniques*, PACT '07, pages 49–59, Washington, DC, USA, 2007. IEEE Computer Society. DOI: 10.1109/PACT.2007.4336199 Cited on page(s) 131

[239] L. Wang, J. M. Stichnoth, and S. Chatterjee. Runtime performance of parallel array assignment: an empirical study. In *Proceedings of the 1996 ACM/IEEE conference on Supercomputing (CDROM)*, Supercomputing '96, Washington, DC, USA, 1996. IEEE Computer Society. DOI: 10.1145/369028.369036 Cited on page(s) 101, 128

[240] M. N. Wegman and F. K. Zadeck. Constant propagation with conditional branches. *ACM Trans. Program. Lang. Syst.*, 13(2):181–210, 1991. DOI: 10.1145/103135.103136 Cited on page(s) 14, 125

[241] M. Weiser. Program slicing. In *Proceedings of the 5th International Conference on Software engineering*, pages 439–449, Piscataway, NJ, USA, 1981. IEEE Press. DOI: 10.1109/FOSM.2008.4659249 Cited on page(s) 34, 130

[242] M. E. Wolf and M. Lam. A data locality optimizing algorithm. In *ACM Conference on Programming Language Design and Implementation*, pages 30 – 44, 1991. DOI: 10.1145/113446.113449 Cited on page(s) 93, 94, 127

[243] M. E. Wolf and M. Lam. Loop transformation and algorithm to maximize parallelism. *IEEE Transaction on Parallel and Distributed Computing*, 2(4):452 – 471, October 1991. DOI: 10.1109/71.97902 Cited on page(s) 93, 94, 127

[244] M. Wolfe. Loop skewing: the wavefront method revisited. *Int. J. Parallel Program.*, 15:279–293, October 1986. DOI: 10.1007/BF01407876 Cited on page(s) 71, 126, 127

[245] M. Wolfe. Iteration space tiling for memory hierarchies. In *Proceedings of the Third SIAM Conference on Parallel Processing for Scientific Computing, Los Angeles, California, USA, December 1-4, 1987*, pages 357–361. SIAM, 1987. Cited on page(s) 127

[246] M. Wolfe. Multiprocessor synchronization for concurrent loops. *IEEE Software*, 5(1):34–42, 1988. DOI: 10.1109/52.1992 Cited on page(s) 126

[247] M. Wolfe. Vector optimization vs vectorization. *J. Parallel Distrib. Comput.*, 5(5):551–567, 1988. DOI: 10.1016/0743-7315(88)90012-3 Cited on page(s) 126

[248] M. Wolfe. Beyond induction variables. In *Proceedings of the ACM SIGPLAN 1992 conference on Programming language design and implementation*, pages 162–174, New York, NY, USA, 1992. ACM. DOI: 10.1145/143095.143131 Cited on page(s) 69, 127

[249] M. Wolfe. The definition of dependence distance. *ACM Trans. Program. Lang. Syst.*, 16(4):1114–1116, 1994. DOI: 10.1145/183432.183440 Cited on page(s) 125

[250] M. Wolfe. *High performance compilers for parallel computing*. Addison-Wesley Publishing Company, 1996. Cited on page(s) 71, 84, 126, 127

[251] M. Wolfe. Parallelizing compilers. *ACM Comput. Surv.*, 28:261–262, March 1996. DOI: 10.1145/234313.234417 Cited on page(s) 125, 126

[252] M. Wolfe and C. W. Tseng. The Power Test for data dependence. *IEEE Trans. Parallel Distrib. Syst.*, 3:591–601, September 1992. DOI: 10.1109/71.159042 Cited on page(s) 37, 126

[253] P. Wu, A. Cohen, and D. Padua. Induction variable analysis without idiom recognition: beyond monotonicity. In *Proceedings of the 14th International Conference on Languages and compilers for parallel computing*, pages 427–441, Berlin, Heidelberg, 2003. Springer-Verlag. DOI: 10.1007/3-540-35767-X_28 Cited on page(s) 127

[254] K. Yotov, X. Li, G. Ren, M. Cibulskis, G. DeJong, M. J. Garzarán, D. A. Padua, K. Pingali, P. Stodghill, and P. Wu. A comparison of empirical and model-driven optimization. In *Proceedings of the ACM Conference on Programming Language Design and Implementation*, pages 63–76, 2003. DOI: 10.1145/781131.781140 Cited on page(s) 128

[255] A. Zhai, C. B. Colohan, J. G. Steffan, and T. C. Mowry. Compiler optimization of memory-resident value communication between speculative threads. In *Proceedings of the International nymposium on Code Generation and Optimization: feedback-directed and runtime optimization*, CGO '04, pages 39–, Washington, DC, USA, 2004. IEEE Computer Society. DOI: 10.1109/CGO.2004.1281662 Cited on page(s) 131

[256] H. Zhong, M. Mehrara, S. A. Lieberman, and S. A. Mahlke. Uncovering hidden loop level parallelism in sequential applications. In *14th International Symposium on High Performance Computer Architecture (HPCA)*, pages 290–301, 2008. DOI: 10.1109/HPCA.2008.4658647 Cited on page(s) 131

[257] W. Zhu, V. C. Sreedhar, Z. Hu, and G. R. Gao. Synchronization state buffer: supporting efficient fine-grain synchronization on many-core architectures. In *Proceedings of the 34th annual International Symposium on Computer Architecture*, ISCA '07, pages 35–45, New York, NY, USA, 2007. ACM. DOI: 10.1145/1273440.1250668 Cited on page(s) 126

[258] C. Zilles and G. Sohi. Execution-based prediction using speculative slices. In *Proceedings of the 28th annual International Symposium on Computer Architecture*, ISCA '01, pages 2–13, New York, NY, USA, 2001. ACM. DOI: 10.1145/379240.379246 Cited on page(s) 130

[259] C. Zilles and G. Sohi. Master/slave speculative parallelization. In *Proceedings of the 35th annual ACM/IEEE International Symposium on Microarchitecture*, MICRO 35, pages 85–96, Los Alamitos, CA, USA, 2002. IEEE Computer Society Press. DOI: 10.1109/MICRO.2002.1176241 Cited on page(s) 130

Author's Biography

SAMUEL P. MIDKIFF

Samuel Midkiff (https://engineering.purdue.edu/~smidkiff/) is a Professor of Electrical and Computer Engineering at Purdue University, where he has been since 2001. He received his PhD degree from the University of Illinois at Urbana-Champaign in 1992 where he was a member of the Cedar project. In 1991 he became a Research Staff Member at the IBM T.J. Watson Research Center, where he was a key member of the xlhpf compiler team and the Numerically INtensive Java (Ninja) project. His research has focused on parallelism and high performance computing, and in particular compiler and language support for the development of correct and efficient programs. To this end, his research has covered dependence analysis and automatic synchronization of explicitly parallel programs, compilation under different memory models, automatic parallelization, high performance computing in Java and other high-level languages, and tools to help in the detection and localization of program errors.

Printed in the United States
by Baker & Taylor Publisher Services